Student Study Guide/ Solutions Manual

to accompany

Biochemistry: An Introduction

Second Edition

Dr. Trudy McKee
Thomas Jefferson University

Dr. James R. McKee
University of the Sciences in Philadelphia

Prepared by
Bruce Morimoto
Purdue University

Boston   Burr Ridge, IL   Dubuque, IA   Madison, WI   New York   San Francisco   St. Louis
Bangkok   Bogotá   Caracas   Lisbon   London   Madrid
Mexico City   Milan   New Delhi   Seoul   Singapore   Sydney   Taipei   Toronto

WCB/McGraw-Hill

A Division of The **McGraw·Hill** *Companies*

Student Study Guide/Solutions Manual to accompany
BIOCHEMISTRY: AN INTRODUCTION, SECOND EDITION

This book is printed on acid-free paper.

    5 6 7 8 9 BKM BKM 9 0 9 8 7 6 5 4 3 2 1

ISBN 0-07-290503-4

www.mhhe.com

Contents

The chemistry of living organisms is immensely complex. Every day, scientists are making strides in understanding how chemical reactions are organized and regulated. Despite the seemingly limitless amount of biochemical information which exists today, there are some fundamental concepts and principles which this study guide will highlight.

General Study Tips

A combination of approaches is necessary to effectively learn biochemistry. Memorization of basic concepts and structures, for example, is important. But memorization alone will only take you so far. Biochemistry is akin to learning a foreign language: you must first learn the vocabulary and grammar before you will be able to communicate. Biochemists have developed their own "vocabulary," such as the names and structures of amino acids, nucleic acids, sugars, and lipids. Biochemical "grammar," so to speak, involves chemical reactions which transform one molecule into another. The immensity of biochemical pathways, reactions, and compounds makes it unreasonable for anyone to memorize everything.

One important commonality in basic biochemistry is that there is often a function or purpose for every metabolic pathway which occurs. Although some of these functions may not be immediately obvious or understood, it is nonetheless critically important to learn and remember them. In studying biochemistry, be sure to think about the function of reactions. Why do they occur? What is their significance? In asking questions, however, be mindful that biochemistry is a dynamic and constantly changing discipline which does not have all the answers. The excitement of biochemical research comes in pursuing and finding answers to many challenging questions.

Special Thanks

I must acknowledge and thank all of my former instructors for their excitement and enthusiasm about science, which in turn stimulated my interest in biochemistry. These individuals include UCLA professors Steven Clarke, Richard Weiss, Charles West, and, most importantly, my Ph.D. advisor Daniel Atkinson. I would also like to thank Mike Fester, my introductory biochemistry teaching assistant, for his encouragement in my entering the world of academia, my wife Joy for her tireless support, and Joel Dirksen, who assisted me in compiling information for this study guide.

Best wishes in your studies of biochemistry!

Bruce H. Morimoto
March 1998

1 Biochemistry: An Introduction

Objectives

1. What principles are central to our understanding of living organisms?

    The following principles have been established by biochemists to be central to our understanding of living organisms:

    a. Cells, the basic structural units of all living organisms, are highly organized. A constant source of energy is required for the maintenance of a cell's ordered state.

    b. Living processes consist of thousands of chemical reactions. Precise regulation and integration of these reactions are required for maintenance of life.

    c. Certain fundamental reaction pathways, such as the energy-generating conversion of glucose to pyruvic acid, known as glycolysis, are found in all organisms.

    d. All organisms utilize the same types of molecules. Examples of such biomolecules include carbohydrates, lipids, proteins, and nucleic acids.

    e. Encoded in each organism's nucleic acid are instructions for growth, development, and reproduction.

2. What characteristics distinguish prokaryotes from eukaryotes?

    Prokaryotes lack a true nucleus, unlike eukaryotes which contain a complex membrane-bound structure of genetic information. Along with well-formed nuclei, eukaryotic cells contain a number of subcellular structures called organelles. Given the simplicity of prokaryotic cells, biochemical diversity has allowed various species to occupy not only all temperate environments, but also harsh and seemingly lifeless ones. One of the most noticeable differences between the two cells is that eukaryotic cells are much larger than prokaryotic cells.

3. What are the four major types of small biomolecules found in cells?

    Cells contain four families of small biomolecules: amino acids, sugars, fatty acids, and nucleotides.

4. What are the primary functions of metabolism in living organisms?

    Metabolism – the total chemical reactions in a cell – performs the primary function of acquisition and utilization of energy, synthesis of molecules needed for cell structure and functioning, and removal of waste.

5. What are the most common types of chemical reactions in living organisms?

    Among the more frequent reactions encountered in biochemical processes are the following: nucleophilic substitution reactions, elimination, isomerization, oxidation-reduction, and hydrolysis.

6. How do cells maintain a high degree of internal order?

    The complex structure of cells requires a high degree of internal order. This is accomplished in four ways: synthesis of biomolecules, transport of ions and molecules across membranes, production of force and movement, and the removal of metabolic waste products and other potentially toxic substances.

General Principles

Biochemistry is the study of how living cells form and break apart molecules. The function of the cell is: 1) to live, which requires energy (light for plants, and chemical for animals) and, 2) to grow, which requires synthesis of new biomolecules.

How do cells satisfy these requirements?

Plants convert light energy into chemical energy (**photosynthesis**). Animals then consume plants as a source of chemical energy in which they convert larger molecules into smaller ones (**catabolism**), and in the process generate energy (in the form of ATP, the energy currency of the cell). Using this energy, the smaller molecules can be converted into larger molecules (**anabolism**).

Catabolism

Living cells extract chemical energy by breaking down molecules. Energy is in the form of electrons. The removal of electrons from a molecule is called **oxidation**.

$$C_6H_{12}O_6 \longrightarrow CO_2 + H_2O + electrons$$

The carbon in sugar loses electrons and is therefore oxidized. The reverse of oxidation is called **reduction**.

$$CO_2 \longrightarrow C_6H_{12}O_6$$

In order for something to be oxidized, another molecule must be simultaneously reduced.

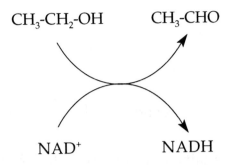

$$CH_3\text{-}CH_2\text{-}OH \qquad CH_3\text{-}CHO$$

$$NAD^+ \qquad NADH$$

Ethanol, for example, loses electrons and is therefore oxidized. NAD^+ accepts electrons and is therefore reduced.

Catabolism can be either:
    Aerobic (with air, or oxygen) or
    Anaerobic (without air), sometimes referred to as fermentation

Plants, animals, and bacteria catabolize sugars to produce energy. Six-carbon sugars are broken down into three-carbon molecules, thereby generating some energy and electrons. Without oxygen, the electrons have no where to go, and the molecules which carry or transfer electrons cannot be recycled.

In yeast and some bacteria, the electrons are donated back to a three-carbon molecule to form ethanol and carbon dioxide. This is called **fermentation** and is the basis by which yeast are used to make beer and wine (ethanol), or make breads rise (production of CO_2).

For animals, in the absence of oxygen, a three-carbon molecule accepts the electrons and is converted into lactic acid. As the name denotes, lactic acid is an acid and is responsible for muscle cramps. When we exercise vigorously, for example, our muscles become starved for oxygen. The electron-carrying intermediates recycle and form lactic acid. This results in a drop in muscle pH, which in turn results in a muscle cramp.

What does oxygen do?
It allows three-carbon molecules to be oxidized to CO_2, because the electrons are transferred to oxygen, converting oxygen into water. Oxygen functions as a terminal electron acceptor.

2   Living Cells

Objectives

1. How are prokaryotes and eukaryotes alike and how are they different?

    Similarities

    Both cells are self-contained units surrounded by a membrane that separates them from their environment.

    The two cells are composed of the same type of molecules.

    Differences

    The most obvious difference between the two cells is their size: the eukaryotic cell is much larger than the prokaryotic cell.

    Another distinct difference is that the eukaryotic cell has a complex internal structure with a variety of membrane-bound organelles, whereas the prokaryotic cell has simple structures.

2. Which organelles carry out the processes required to maintain a living state in eukaryotic cells?

    A eukaryotic cell is composed of seven basic organelles which maintain its living state:

    > Plasma membrane: Performs selective permeability in which certain substances are prevented from entering the cell and others from leaking out.

    > Nucleus: Contains the cell's information; exerts a profound influence over all cellular metabolic activities; plays a major role in the synthesis of the RNA components of ribosomes.

    > Endoplasmic reticulum: Primarily involved in protein synthesis, lipid synthesis, and biotransformation.

    > Cytoplasmic ribosomes: Involved in the biosynthesis of proteins.

    > Peroxisomes: Noted for their involvement in the generation and breakdown of peroxides.

    > Mitochondria: Involved in the biosynthesis of ATP through aerobic respiration.

    > Cytoskeleton: Involved in the maintenance of cell shape, facilitation of cellular movement, and the intracellular transport of organelles.

3. How might eukaryotic cells have evolved from prokaryotic cells?

According to the endosymbiotic hypothesis, eukaryotic cells began as large anaerobic organisms. Mitochondria arose when small aerobic bacteria were ingested by the larger cells. In exchange for the benefits of protection and a constant nutrient supply, the smaller cell provided its host with energy generated by a process known as **aerobic respiration**. As time passed, the bacteria lost their independence because of the transfer of several genes to the nucleus. Similarly, chloroplasts are believed to have descended from cells which were similar to modern cyanobacteria, while cilia and flagella were derived from ancient spiral prokaryotes.

4. What technologies have aided scientists in discovering how cells function?

Much of the current knowledge of biochemical processes is due directly to technological innovations. There are several technologies which have had an enormous impact on biochemistry, such as cell fractionation, the electron microscope, and autoradiography. Cell fractionation techniques allow the study of cell organelles in a relatively intact form outside of cells. The electron microscope provides a view of cellular ultrastructure that is not possible with the use of the more commonly available light microscope. Autoradiography is a technique that is used to study the intracellular location and behavior of cellular components.

General Principles

Chemical reactions do not occur randomly in the cell; there exists quite a bit of organization. This biological organization forms two major classifications of organisms:

Prokaryotic (meaning, before the nucleus)
Eukaryotic (meaning, true nucleus)

Bacteria are the best example of prokaryotes. They are single-celled organisms and perhaps the most adaptable creatures on earth. They are found almost everywhere and in extreme temperatures ranging from glaciers in Antarctica to undersea volcanoes in the Pacific Ocean. Bacteria can perform nearly any chemical reaction; from consuming oils, for instance, to producing pharmaceuticals.

Eukaryotes differ from prokaryotes in that eukaryotes have organelles.

Organelles are compartments within the cell that have specialized functions. The compartments are separated from the rest of the cell by membranes which function as barriers.

3   Water: The Medium of Life

Objectives

1. How is the molecular structure of water related to its chemical behavior?

    Water is composed of two atoms of hydrogen and one atom of oxygen. Each hydrogen atom is linked to the oxygen atom by a single covalent bond. Water molecules have a bent geometry with a bond angle of 104.5°. Because the oxygen atom is more electronegative than the hydrogen atom, the unequal electron distribution results in a bent geometry and polarity of the molecule. The chemical behavior of water is largely dependent upon its polarity and the intermolecular interactions which can occur.

2. How does non-covalent bonding affect the chemical and biological properties of water?

    There are four basic non-covalent bonding interactions which determine the capacity of water to interact with other types of molecules. These are hydrogen bonding, electrostatic interactions, van der Waal's forces, and hydrophobic interactions. These interactions are especially important because biological reactions take place in a water medium and determine the shape and function of biomolecules.

    Hydrogen bonds: One consequence of the large difference in electronegativity between hydrogen and oxygen is that the hydrogen of one water molecule is attracted to the unshared pairs of electrons of another water molecule, thereby forming a hydrogen bond. The hydrogen bond is largely responsible for the thermal properties of water.

    Electrostatic interactions: These interactions occur between oppositely charged atoms or groups. This is an important aspect because electrostatic interactions are responsible for the hydration of ions. The polar water molecules form shells of water molecules when they are attracted to the charged ions. As ions become hydrated, the attractive force between them is reduced and the charged species dissolve in the water.

    Van der Waal's forces: This is a class of weak, transient electrostatic interactions which occur between permanent and/or induced dipoles. They may be attractive or repulsive, depending on the distance between the atoms or groups involved.

Hydrophobic interactions: Hydrophobic interactions occur when small amounts of non-polar substances are observed to coalesce into droplets when mixed with water. Hydrophobic interactions have a profound effect on living cells. For example, they are responsible for the structure of membranes and the stability of proteins.

3.   What is pH and how does it affect living cells?

pH is defined as the negative logarithm of the concentration of hydrogen ions.

The concentration of hydrogen ions affects most cellular and organismal processes. For example, the structure and function of proteins and the rates of most biochemical reactions are strongly affected by hydrogen ion concentration. Additionally, hydrogen ions play a major role in processes such as energy generation and endocytosis.

4.   What is a buffer? What role do buffers play in living cells?

A buffer is a solution which contains a weak acid and its salt, and is resistant to large pH changes upon addition of stronger acids or bases. The three most important buffers in the body are: bicarbonate, phosphate, and protein buffers. Bicarbonate controls the pH in the blood; phosphate buffer controls the pH of the intracellular fluids; and protein buffer controls the pH of the blood and many other fluids.

5.   What are the colligative properties of water solutions?

Colligative properties (vapor pressure depression, boiling point elevation, freezing point depression, and osmotic pressure) are grouped together because they depend on the number of dissolved particles in a given mass solvent rather than the identity of the particles.

General Principles

Approximately 70-80% of our body weight is water. Water is essential for life as we know it. Water has unique chemical properties, some of which allow biomolecules to function

What makes water unique?

The chemical nature of water confers some of these special properties:

A **covalent bond** is formed between two atoms when those atoms share electrons between each other. When the atoms share the electrons equally, the electron spends the same amount of time around each atom. This gives rise to a **non-polar** bond.

7

Some atoms are more attractive to electrons than others. This property is called **electronegativity**. The electronegativity of common biological atoms is $H \approx C < N < O$ (order of increasing electronegativity in which oxygen is the most electronegative atom of the series).

For a C-O bond, the electrons will spend more time around oxygen than carbon. When hydrogen is bonded to either nitrogen or oxygen, the electrons will be more attracted to oxygen or nitrogen.

Electrons are negatively charged, which gives rise to partial charges denoted as δ (Greek lowercase delta). The unequally shared electrons give rise to a **polar** bond.

While a particular bond can be polar, overall molecules can still be non-polar. This is because geometry also influences the polarity of a molecule.

<u>Example</u>: CO_2

$$O=C=O$$

Each C=O bond is polar, but since CO_2 is linear, the polar bond of each half cancels out the other, and the molecule CO_2 is therefore non-polar.

Since water is bent at an 105° angle, the direction of the polar bonds cannot cancel one another out. This gives rise to a **dipole**, which means that one end has a δ^- and the other has a δ^+.

The polar nature of water allows it to interact with a variety of other molecules. For example, table salt (NaCl) dissolves completely in water because of ion-dipole interactions. Alcohol dissolves in water via dipole-dipole interactions.

This gives rise to two terms:

<u>Hydrophilic</u>: meaning water-loving, such as polar and ionic molecules,

and

<u>Hydrophobic</u>: meaning water-fearing, such as non-polar molecules

Molecules can have both polar and non-polar ends; these molecules are termed **amphiphilic**. For example, salts of fatty acids are soaps. Soap was first produced by boiling animal fat with lye (NaOH). Sodium lauryl sulfate, a synthetic soap, is found in many household products such as toothpaste.

Why are soaps effective at getting rid of grease?

The hydrophobic tail interacts with the non-polar "grease" molecule. This allows the polar end of the soap to interact with water. Water is then able to completely surround the grease molecule, allowing the grease/detergent micelle to become water-soluble.

Study Tips

<u>Important Concepts</u>

- polarity of chemical bonds

- non-covalent bonds

- pH

- acids/bases

<u>How to solve acid-base problems</u>

All problems use the Henderson-Hasselbalch equation:

$$pH = pK_a + \log \frac{[base]}{[acid]}$$

The single greatest hurdle in solving acid-base problems is identifying the weak acid and its conjugate base.

Remember that an acid is a proton donor and a base is a proton acceptor; the weak acid will therefore have an extra proton when compared to its conjugate base.

<u>Example</u>:

HCO_3^-	CO_3^{2-}
acid	conjugate base

$H_2PO_4^-$	HPO_4^{2-}
acid	conjugate base

When asked, "What is the pH of...?" you will want to solve for pH. You will therefore need to have the pK_a and the [acid] and [base].

When asked, "What is the charge of...?" you will want to solve for the ratio:

$$\frac{[base]}{[acid]}$$

Remember that the acid HA, and the base A⁻, will differ by **one** charge. For example, if a molecule has a carboxyl group as the weak acid,

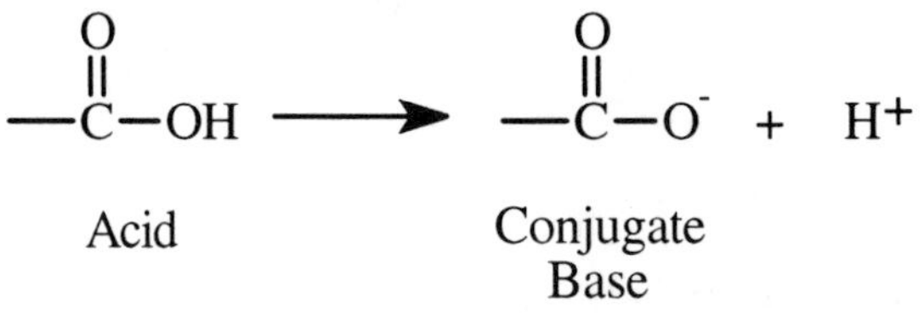

$$\underset{\text{Acid}}{-\overset{\overset{\displaystyle O}{\|}}{C}-OH} \longrightarrow \underset{\substack{\text{Conjugate} \\ \text{Base}}}{-\overset{\overset{\displaystyle O}{\|}}{C}-O^-} \; + \; H^+$$

When the ratio $\dfrac{[A^-]}{[HA]}$ is small, the molecule will be uncharged.

When the ratio $\dfrac{[A^-]}{[HA]}$ is large, the molecule will be negatively charged.

Examples:

1. What is the pH of a phosphate buffer when the $[H_2PO_4^-] = 10$ mM and the $[HPO_4^{2-}] = 1$ mM? The $pK_a = 6.7$

    First identify the acid and the conjugate base.

    $$H_2PO_4^- \longrightarrow HPO_4^{2-} + H^+$$

    Since $H_2PO_4^-$ donates a protein, it is the acid and HPO_4^{2-} is the conjugate base.

    The Henderson-Hasselbalch equation, $pH = pK_a + \log \dfrac{[\text{base}]}{[\text{acid}]}$ is then used to determine the pH of the buffer solution.

    $$pH = 6.7 + \log \dfrac{[HPO_4^{2-}]}{[H_2PO_4^-]} = 6.7 + \log \dfrac{1 \text{ mM}}{10 \text{ mM}} = 6.7 + \log 0.1 = 5.7$$

2. Calculate the ratio of $\dfrac{[HPO_4^{2-}]}{[H_2PO_4^-]}$ at pH 5.7, 6.7, and 7.7. The pK_a is 6.7.

    Again, first identify the acid and the conjugate base.

    $$H_2PO_4^- \longrightarrow HPO_4^{2-} + H^+$$

    As before, $H_2PO_4^-$ is the acid and HPO_4^{2-} is the conjugate base.

    The Henderson-Hasselbalch equation, $pH = pKa + \log \dfrac{[\text{base}]}{[\text{acid}]}$ is then rearranged to solve for the ratio, $\dfrac{[\text{base}]}{[\text{acid}]}$.

    $$\log \dfrac{[\text{base}]}{[\text{acid}]} = pH - pK_a$$

$$\frac{[\text{base}]}{[\text{acid}]} = 10^{(\text{pH} - pK_a)}$$

At pH 5.7, $\dfrac{[\text{base}]}{[\text{acid}]} = 10^{(5.7 - 6.7)} = 10^{-1} = 0.1$ or $\dfrac{1}{10}$

At pH 6.7, $\dfrac{[\text{base}]}{[\text{acid}]} = 10^{(6.7 - 6.7)} = 10^{0} = 1$ or $\dfrac{1}{1}$

At pH 7.7, $\dfrac{[\text{base}]}{[\text{acid}]} = 10^{(7.7 - 6.7)} = 10^{1} = 10$ or $\dfrac{10}{1}$

3. What is the concentration of CO_2 in blood when the pH of blood is 7.1 and the concentration of HCO_3^- is 8 mM? The pK_a is 6.1.

First identify the acid and the conjugate base.

$$CO_2 + H_2O \longrightarrow H_2CO_3 \longrightarrow HCO_3^- + H^+$$

CO_2 is the acid and HCO_3^- is the conjugate base. Let us now rearrange the Henderson-Hasselbalch equation to solve for the concentration of acid (CO_2).

$$pH = pKa + \log \frac{[\text{base}]}{[\text{acid}]}$$

$$\log \frac{[\text{base}]}{[\text{acid}]} = pH - pK_a$$

$$\frac{[\text{base}]}{[\text{acid}]} = 10^{(\text{pH} - pK_a)}$$

$$[\text{acid}] = \frac{[\text{base}]}{10^{(\text{pH} - pK_a)}}$$

$$[CO_2] = \frac{[HCO_3^-]}{10^{(\text{pH} - pK_a)}}$$

$$[CO_2] = \frac{8 \text{ mM}}{10^{(7.1 - 6.1)}} = 0.8 \text{ mM}$$

11

4. What is the concentration of HCO_3^- and CO_2 at pH 7.4 when the concentration of the buffer is 25.2 mM?

From the previous example we know that CO_2 is the acid and HCO_3^- is the conjugate base, and the pK_a is 6.1.

$$\frac{[HCO_3^-]}{[CO_2]} = 10^{(pH - pK_a)} = 10^{(7.4 - 6.1)} = 10^{1.3} = 20$$

The concentration of buffer means, $[CO_2] + [HCO_3^-] = 25.2$ mM, and since the Henderson-Hasselbalch equation told us that $\frac{[HCO_3^-]}{[CO_2]} = 20$

$$[CO_2] + 20\,[CO_2] = 25.2 \text{ mM}$$

$$21\,[CO_2] = 25.2 \text{ mM}$$

$$[CO_2] = 1.2 \text{ mM}$$

$$[HCO_3^-] = 25.2 \text{ mM} - 1.2 \text{ mM} = 24 \text{ mM}$$

4 Energy

Objectives

1. How do the principles of thermodynamics apply to living organisms?

   The principles of thermodynamics involve heat and energy transformations. The study of energy transformations in living cells is called bioenergetics. Bioenergetics focuses on the initial and final states of biomolecules, and helps determine the direction and extent to which specific biochemical reactions take place.

2. How is thermodynamics used to determine whether specific biochemical reactions are spontaneous?

   The second law of thermodynamics states that any system, no matter how organized it may be, tends to become increasingly disorganized. In any spontaneous process there is always an increase in entropy. To predict whether a given process is spontaneous, the sign of the ΔS must be known. For example, if the value of ΔS is positive, then the process is spontaneous. If the ΔS is negative, then the process will not occur as expected. In fact, the reverse process would take place. If ΔS is equal to zero, the process has no tendency to occur. For instance, organisms which are at equilibrium with their surroundings are dead.

3. What is the significance of free energy?

   Free energy is a state function that relates the first and second laws of thermodynamics in the expression: $\Delta G = \Delta H - T\Delta S$, where ΔG is Gibbs free energy, ΔH is the enthalpy change, T is the Kelvin temperature, and ΔS is the entropy change. Free energy represents the maximum amount of work obtained from a process. If a process occurs at a constant temperature, pressure, and volume, it will be spontaneous in the direction in which free energy decreases. Such processes are called **exergonic**. If the free energy change is positive, the process is called **endergonic**. When the free energy change in a system is equal to zero, the system is at **equilibrium**.

4. How are oxidation-reduction reactions involved in the generation of energy?

   In living organisms, both energy-capturing and energy-releasing processes consist primarily of redox reactions. Redox reactions occur when there is a transfer of electrons between an electron donor and electron acceptor.

5.   What is the role of ATP in living systems?

Adenosine triphosphate (ATP) plays an extraordinary role within living cells. The hydrolysis of ATP provides an immediate and direct input of free energy to drive an immense variety of endergonic biochemical reactions. These reactions include synthesis of biomolecules, active transport of substances across cell membranes, and mechanical work such as muscle contraction.

General Principles

Thermodynamics is the relationship between different forms of energy. Thermodynamics allows us to predict whether a particular chemical reaction can occur. It is important to remember that thermodynamics provides us with an understanding as to whether a particular chemical conversion is possible, not whether it will actually occur.

Example:

Glucose can be converted to carbon dioxide and water in the presence of water. The thermodynamics of this reaction tells us that this can happen (and of course it does in our cells). However, the sugar in a bowl in our kitchen, for example, does not automatically oxidize to carbon dioxide because the kinetics of the uncatalyzed reaction is very slow.

Thermodynamic properties are "state" dependent or "state" functions, meaning they do not depend on how a substance was made or how it reached that state of being. In thermodynamics we are only interested in the difference between the initial and the final points – it is not important how you get from point A to point B.

The quantity known as **free energy** (abbreviated "G") allows one to predict whether a chemical reaction can occur or not. For our purposes, we are only interested in relative changes, known as ΔG.

The sign of ΔG is the predictive element.

$-\Delta G$ $\longrightarrow$ reaction favorable (exergonic, spontaneous)

$+\Delta G$ $\longrightarrow$ reaction not favorable (endergonic, non-spontaneous)

$\Delta G = 0$ $\longrightarrow$ reaction at equilibrium (no change)

Since we are only interested in changes in free energy, there needs to be some reference point. For example, geographic elevations may be described as 10,000 feet **above sea level**. Sea level is our point of reference. The reference point for free energy is given as $\Delta G°$, called the **standard free energy**. The standard state is defined as 25°C, 1 atm pressure, and 1 M reactant concentration. Nearly all biochemical reactions, however, occur in dilute, aqueous mixtures, so biochemists have developed their own reference point known as $\Delta G°'$, the modified standard free energy under biological conditions. $\Delta G°$ and $\Delta G°'$ are related by the following equation:

$$\Delta G°' = \Delta G° + 2.303\ RT \log [H_2O],$$

where the $[H_2O]$ is approximately equal to 55.5 M for dilute solutions.

ΔG, in an actual chemical reaction, is related to the reference $\Delta G°$ (or $\Delta G°'$) by the following equation:

$$\Delta G = \Delta G° + 2.303\ RT \log \frac{[products]}{[reactants]}$$

Consider a simple reaction in which A $\longrightarrow$ B. If the ΔG for this reaction is negative (i.e., favorable), then the reverse reaction in which B $\longrightarrow$ A has to be unfavorable ($+\Delta G$). Since ΔG is a state function, the magnitude of ΔG in forward and reverse reactions is the same. So when reversing the direction of a chemical reaction, one need only to change the sign of ΔG.

When ΔG is equal to zero, there is no net change in the chemical reaction, meaning that the rate of formation precisely equals the rate of reversal. This is a condition in which the reaction is at equilibrium. When $\Delta G = 0$,

then $\Delta G° = -\ 2.303\ RT \log \dfrac{[products]_{at\ equilibrium}}{[reactants]_{at\ equilibrium}}$

$\Delta G°$ is therefore related to the equilibrium constant of a reaction by the following equation:

$$\Delta G° = -\ 2.303\ RT \log K_{eq}$$

In the $\Delta G = \Delta G° + 2.303\ RT \log \dfrac{[products]}{[reactants]}$ equation, the $RT \log\{\ \}$ quantity describes how changing the concentration of products or reactants can alter the direction of a chemical reaction.

Biological harnessing of energy

Three requirements for life are:

- Energy

- Reducing power (source of electrons)

- Chemical intermediates

Catabolism produces:

- Energy (as ATP)

- Electrons (NADH or NADPH)

- Intermediates

For the macromolecular synthesis of proteins, DNA, and RNA, anabolism requires:

- Energy (as ATP)

- Electrons (NADH or NADPH)

- Intermediates

Chemical coupling allows the cell to harness the energy produced by catabolism. You may want to think of a chemical reaction which has negative ΔG as a flowing river. The water is moving downstream, just as a chemical reaction is moving in the forward direction. The water has potential energy to do work. If we were to construct a dam in the river, the flowing water could turn turbines to generate electricity. Similarly, a chemical reaction proceeding in the forward direction can be used to drive some other chemical reaction. This is the basis of chemical coupling.

Example:

$$\text{Phosphoenolpyruvate (PEP) + water} \longrightarrow \text{pyruvate} + P_i$$

$\Delta G^{\circ\prime} = -14.8 \dfrac{\text{kcal}}{\text{mol}}$, therefore this reaction is favorable (or as written, will be moving in the forward direction). To capture some of the energy of this reaction, consider a thermodynamically unfavorable reaction (one with a $+ \Delta G^{\circ\prime}$) such as:

$$\text{ADP} + P_i \longrightarrow \text{ATP + water}$$

The $\Delta G^{\circ\prime}$ is $+ 7.3 \dfrac{\text{kcal}}{\text{mol}}$ for this reaction. If we sum the two reactions:

$$\text{PEP + water} \longrightarrow \text{pyruvate} + P_i$$
$$\underline{\text{ADP} + P_i \longrightarrow \text{ATP + water}}$$
$$\text{PEP + ADP} \longrightarrow \text{pyruvate + ATP}$$

The $\Delta G^{\circ\prime}$ of the overall reaction is simply the sum of $\Delta G^{\circ\prime}$ of the individual reactions: $- 7.5 \dfrac{\text{kcal}}{\text{mol}}$ ($- 14.8 \dfrac{\text{kcal}}{\text{mol}} + 7.3 \dfrac{\text{kcal}}{\text{mol}}$). The new "coupled" reaction is still thermodynamically favorable (-ΔG°'), but we used some of the free energy to "drive" an unfavorable chemical reaction.

Cells have evolved to use ATP synthesis and hydrolysis as a means of storing and using chemical energy. For that reason, ATP is often considered the "energy currency" of the cell.

ATP hydrolysis (ATP $\longrightarrow$ ADP + P_i) has a ΔG°' of - 7.3 kcal/mol, and can be used to make an unfavorable chemical reaction favorable. For example,

$$\text{glucose} + P_i \longrightarrow \text{glucose-6-phosphate} + \text{water}$$

has a ΔG°' of + 3.3 $\frac{\text{kcal}}{\text{mol}}$. When this reaction is coupled to ATP hydrolysis, the overall ΔG°' is then - 4 $\frac{\text{kcal}}{\text{mol}}$.

$$\text{glucose} + P_i \longrightarrow \text{glucose-6-phosphate} + \text{water}$$
$$\underline{\text{ATP} + \text{water} \longrightarrow \text{ADP} + P_i}$$
$$\text{glucose} + \text{ATP} \longrightarrow \text{glucose-6-phosphate} + \text{ADP}$$

Study Tips

The sign and value of ΔG° indicates the direction and magnitude of a particular reaction at equilibrium. At equilibrium, ΔG=0, therefore

$$\Delta G° = -2.303 \, RT \log \frac{[\text{products}]_{\text{at eq}}}{[\text{reactants}]_{\text{at eq}}} = -2.303 \, RT \log K_{eq}$$

The equilibrium constant for a reaction is:

$$K_{eq} = \frac{[\text{products}]_{\text{at eq}}}{[\text{reactants}]_{\text{at eq}}}$$

For the hydrolysis of ATP $\longrightarrow$ ADP + P_i, the ΔG°= -7.3 $\frac{\text{kcal}}{\text{mol}}$. The equilibrium constant for this reaction is: $K_{eq} = \frac{[\text{products}]_{\text{at eq}}}{[\text{reactants}]_{\text{at eq}}} = \frac{[\text{ADP}][P_i]}{[\text{ATP}]}$

$$\Delta G° = - 2.303 \, RT \log K_{eq}$$

$$-7.3 \, \frac{\text{kcal}}{\text{mol}} = -2.303 \, RT \log K_{eq}$$

$$- 7300 \, \frac{\text{cal}}{\text{mol}} = -2.303 \, (1.987 \, \frac{\text{cal}}{\text{mol K}}) \, (298 \text{ K}) \log K_{eq}$$

17

$$\frac{-7300 \, \frac{cal}{mol}}{-2.303 \, (1.987 \, \frac{cal}{mol \, K}) \, (298K)} = \log K_{eq}$$

$$5.35 = \log K_{eq}$$

$$10^{5.35} = K_{eq}$$

$$K_{eq} = 2.26 \times 10^5 \, or \, \frac{225,000}{1}$$

The large K_{eq} indicates that the reaction can proceed to the product. In this example, you can see that the negative $\Delta G°$ produced a large K_{eq}. Another way to look at this is that at equilibrium, for every one molecule of ATP, there will be 474 molecules of ADP and P_i, since [ADP] × [P_i] =225,000 and the square root of 225,000 is 474.

What if the $\Delta G°$ was + 7.3 $\frac{kcal}{mol}$?

The K_{eq} would be 4.4×10^6 or $\frac{1}{225,000}$

What does the following mean?

$$ATP \longrightarrow ADP + P_i \qquad\qquad \Delta G° = -7.3 \, \frac{kcal}{mol}$$

and

$$ADP + P_i \longrightarrow ATP \qquad\qquad \Delta G° = + 7.3 \, \frac{kcal}{mol}$$

By reversing the reaction, only the sign of $\Delta G°$ changes. This is an important concept for solving chemical coupling problems.

The actual direction of a reaction can also be altered by changing the concentration of products or reactants. This is described by the general equation:

$$\Delta G = \Delta G° + 2.303 \, RT \log \frac{[products]}{[reactants]}$$

<u>Example:</u>

$$\text{glucose-1-phosphate} \longrightarrow \text{glucose-6-phosphate} \qquad \Delta G^\circ = -1.7 \; \frac{\text{kcal}}{\text{mol}}$$

At equilibrium,

$$\frac{[\text{glucose - 6 - phosphate}]}{[\text{glucose - 1 - phosphate}]} = \frac{17.6}{1}$$

The reaction will proceed to form glucose-6-phosphate. But can this reaction be reversed?

Intuitively, if we were to add a high concentration of glucose-6-phosphate, we would think that this reaction could be "pushed" backwards. What if the concentration of glucose-6-phosphate was 100 mM and the concentration of glucose-1-phosphate was 1 mM? In what direction would the reaction proceed?

Let's use the general equation:

$$\Delta G = \Delta G^\circ + 2.303 \; RT \log \frac{[\text{products}]}{[\text{reactants}]}$$

$$[\text{products}] = [\text{glucose-6-phosphate}] = 100 \; \text{mM}$$

$$[\text{reactants}] = [\text{glucose-1-phosphate}] = 1 \; \text{mM}$$

$$\Delta G = -1700 \; \frac{\text{cal}}{\text{mol}} + 2.303 \; (1.987 \; \frac{\text{cal}}{\text{mol K}}) \; (298 \; \text{K}) \log \frac{100 \; \text{mM}}{1 \; \text{mM}}$$

$$\Delta G = +1030 \; \frac{\text{cal}}{\text{mol}} = +1 \; \frac{\text{kcal}}{\text{mol}}$$

The positive ΔG tells us that this reaction will proceed in the opposite direction as written. So by increasing the concentration of the product, we would be able to reverse the direction of the reaction.

As you can see, the actual direction of a reaction depends not only on the ΔG°, but also on the concentration of reactants and products.

5   Peptides and Proteins

Objectives

1. How does the structure of amino acids affect the proteins which contain them?

    Amino acid structures are responsible for determining the final three-dimensional configuration of each protein. The amino acid classes offer some insight into their role in determining protein structure. Because of their tendency to avoid water, non-polar amino acids play an important role in the maintenance of the three-dimensional structure of proteins. Although the sulfhydryl group of cysteine is non-polar, for example, it does have a slight capacity to form weak hydrogen bonds with oxygen and nitrogen. Sulfhydryl groups, which are highly reactive, are important components of enzymes. Additionally, the sulfhydryl groups of two cysteine molecules may oxidize to form a disulfide bond. Serine, threonine, and tyrosine each contain a polar hydroxyl group which enables them to participate in hydrogen bonding, an important factor in protein structure. The amide functional group is also highly polar, and asparagine and glutamine are capable of hydrogen bonding.

2. In addition to their roles as building blocks, what other functions do amino acids serve?

    In addition to their primary function as components of protein, amino acids have several other biological roles. Several amino acid derivatives act as chemical messengers. Amino acids serve as precursors in the synthesis of a variety of complex nitrogen-containing molecules. And several standard and nonstandard amino acids act as metabolic intermediates.

3. What are peptides and what are some of their functions?

    Peptides are amino acid polymers with low molecular weights, typically consisting of less than fifty amino acids. Some prominent examples are: glutathione, oxytocin, vasopressin, and the opioid peptides. The tripeptide glutathione, found in almost all organisms, is involved in many important biological processes. Among them are protein and DNA synthesis, drug and environmental toxin metabolism, and amino acid transport. Oxytocin stimulates contraction of uterine muscles during childbirth and the ejection of milk by the mammary glands during lactation. In males, oxytocin may have a regulatory role in the synthesis of the sex hormone testosterone.

Vasopressin stimulates the kidneys to retain water, and is secreted in response to low blood pressure or high sodium ion concentrations. Opioid peptides are molecules which relieve pain and produce pleasant sensations.

4.   What type of bonds are important in protein structure?

Three important types of bonds in a protein structure are: peptide bonds, hydrogen bonds, and covalent bonds. Peptide bonds are amide linkages between the α-carboxyl group of one amino acid and the α-amino group of another. The rigidity of the peptide bond has several consequences. Fully one-third of the bonds in a polypeptide backbone chain are unable to rotate freely, which partially accounts for their stability. Another consequence of the peptide bond structure is that successive R groups occur on opposite sides of the polypeptide chain. A significant amount of hydrogen bonding occurs within a protein's interior and on its surface. Many of these hydrogen bonds exist between the carbonyl and the N-H groups in the polypeptide's backbone. The most prominent covalent bonds in a tertiary structure are the disulfide bridges which are found in many extracellular proteins. In extracellular environments these strong linkages protect protein structure to a certain extent from adverse changes in pH or salt concentrations.

5.   What are the four levels of protein structure?

Biochemists have distinguished several levels of the structural organization of proteins. Primary structure – the amino acid sequence – is specified by genetic information. As the polypeptide chain folds, it forms localized arrangements of adjacent amino acids which constitute secondary structure. The final overall, three-dimensional shape that a polypeptide assumes is called the tertiary structure. Proteins which consist of two or more polypeptide chains are said to have a quaternary structure.

6.   What functions do proteins have in living organisms?

Of all of the types of molecules which are encountered in living organisms, proteins have the most diverse functions, as the following list suggests:

Catalysis: Enzymes, the proteins which direct and accelerate thousands of biochemical reactions, are involved in such processes as digestion, energy capture, and biosynthesis.

Structure: Some proteins function as structural materials which provide protection and support.

Movement: Proteins are involved in all types of cellular movement such as cell division, endocytosis, exocytosis, and the ameboid movement of white blood cells.

Defense: A wide variety of proteins have a protective role. For example, the protein found in skin aids in protecting the organism against mechanical and chemical injury; blood clotting proteins prevent blood loss when blood vessels have been damaged; and antibodies serve as protection from invading organisms.

Regulation: The binding of a hormone molecule to a receptor protein on its target cell results in specific changes in cellular function.

Transport: Many proteins function as carriers of molecules or ions across membranes or between cells.

7. What is protein denaturation and how does it occur?

Protein denaturation is the disruption of a protein's native conformation and, depending on the degree of denaturation, the loss of the molecule's biological activity may either be partial or complete. The following conditions can denature a protein: strong acids or bases, organic solvents, detergents, reducing agents, salt concentration, heavy metal, temperature changes, and mechanical stress.

8. What unique properties do fibrous proteins typically have?

As their name suggests, fibrous proteins are insoluble in water and physically resilient. Fibrous proteins typically contain high proportions of regular secondary structures, that is, α-helix or β-pleated sheet. As a consequence of their rod-like or sheet-like shapes, many fibrous proteins have structural rather than dynamic roles. Fibrous proteins, such as the keratins found in skin, hair, and nails, have structural and protective functions.

9. What types of roles do globular proteins play in living organisms?

Globular proteins are compact, spherical molecules which are usually water-soluble. Typically, globular proteins have dynamic functions. Because the function of these proteins usually requires them to bind precisely to other molecules, the identity and arrangement of surface amino acids are important. These residues interact to form specific binding cavities, or clefts, which are complementary to the structure of a specific molecule, called a **ligand**. The remainder of a globular protein's structure is devoted to the binding of regulatory molecules and to ensuring the maintenance of the polypeptide's three-dimensional structure. Nearly all enzymes have globular structures. Other examples of these proteins include immunoglobulins and the transport proteins hemoglobin and albumin.

10. What is allosteric regulation?

Allosteric interactions occur when a small, specific molecule, called an effector or modulator, non-covalently binds to a protein and alters its activity. Any change in activity is due to changes in the interactions among the protein's subunits. Effectors which increase a protein's activity are called **activators**. Those which decrease activity are called **inhibitors**.

Study Tips

Isoelectric point is the pH at which the amino acid has no net change. To determine this you must first:

1. identify the acid and base on that amino acid (what functional groups can be charged?);

2. rank the functional groups in their order of ionization (which groups will be charged and at what pH?);

3. average the pK_as on either side of the isoelectric (neutral) molecule.

Example:

Let's start with alanine:

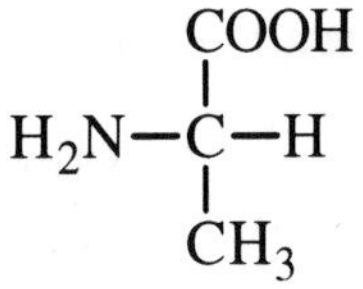

$$
\begin{array}{c}
COOH \\
| \\
H_2N-C-H \\
| \\
CH_3
\end{array}
$$

1. Identify functional groups

   In the physiological pH range (1-14) only -COOH and -NH$_2$ can donate or accept protons, respectively.

   -COOH is a weak acid and can donate a proton. When it does,

   $$-COOH \longrightarrow -COO^- + H^+$$

   the carboxyl group (-COOH) becomes negatively charged (-COO$^-$). This donation of a proton is described by the pK_a, which equals 2.4. Remember, $pK_a = -\log K_a$, and K_a is the equilibrium constant for loss of proton. Likewise, -NH$_2$ is a weak base and can therefore accept a proton and become positively charged (-NH$_3^+$). This occurs with a pK_a of 9.9.

2. Rank functional groups

Begin by asking yourself this: "What will happen to each acid/base group at very low pHs (acidic conditions)?" Both acid and base groups will be protonated. The amino and carboxyl groups will be $-NH_3^+$ and -COOH, so alanine will look like this:

$$
\begin{array}{c}
\text{COOH} \\
| \\
H_3\overset{+}{N}-\text{C}-\text{H} \\
| \\
\text{CH}_3
\end{array}
$$

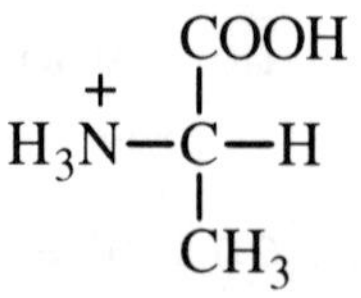

and have a net positive charge of 1.

Consider the opposite extreme: very high pHs (basic conditions). Under these conditions, the acid groups will give up their protons and base groups will have nothing to accept. Both types of groups will be deprotonated and alanine will look like this:

$$
\begin{array}{c}
\text{COO}^- \\
| \\
H_2N-\text{C}-\text{H} \\
| \\
\text{CH}_3
\end{array}
$$

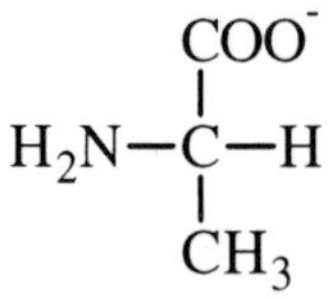

and have a net negative charge of -1.

Now, when going from the all-protonated form to the all-unprotonated form, which group will lose its proton first? The -COOH or the $-NH_3^+$? This is where the pK_as will help us. The group with the lowest pK_a will lose its proton first. Remember, a pK_a is associated with a particular group.

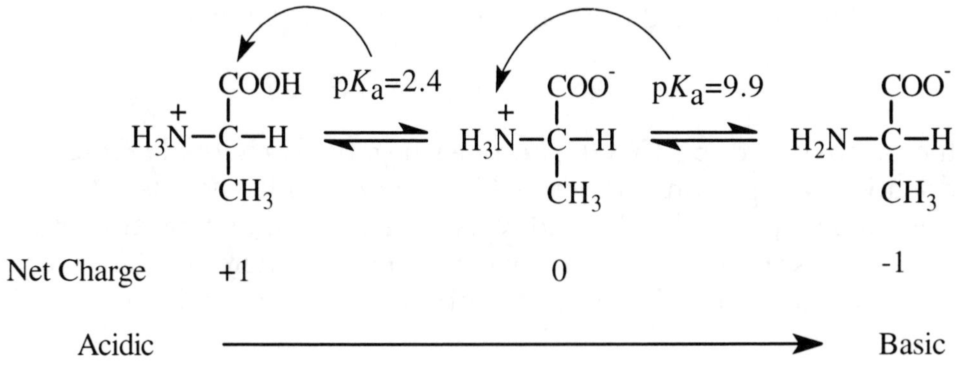

$$
\begin{array}{c}
\text{COOH} \\
| \\
H_3\overset{+}{N}-\text{C}-\text{H} \\
| \\
\text{CH}_3
\end{array}
\quad pK_a=2.4 \quad
\begin{array}{c}
\text{COO}^- \\
| \\
H_3\overset{+}{N}-\text{C}-\text{H} \\
| \\
\text{CH}_3
\end{array}
\quad pK_a=9.9 \quad
\begin{array}{c}
\text{COO}^- \\
| \\
H_2N-\text{C}-\text{H} \\
| \\
\text{CH}_3
\end{array}
$$

Net Charge   +1   0   -1

Acidic ————————————————————————→ Basic

24

3. Average pK_as on either side of the neutral molecule. For alanine this is simple:

$$pI = \frac{2.4 + 9.9}{2} = 7.35$$

Now try this exercise with glutamic acid. There are three functional groups which can be ionized (charged).

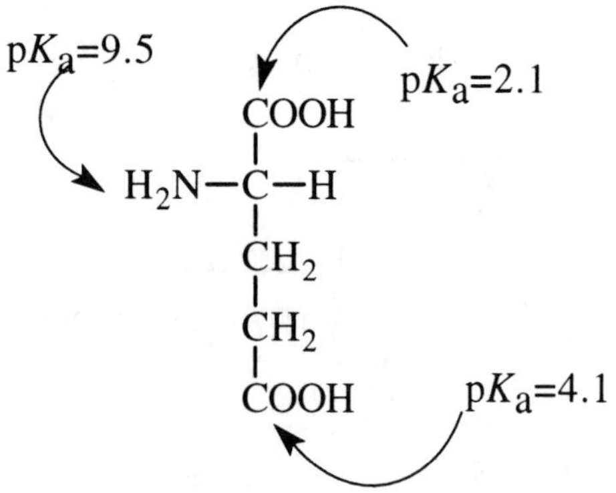

The answer you should get is 3.1 (the average of 2.1 and 4.1).

Peptide Bond

Amino acids are linked by a covalent bond between the α-amino (-NH$_2$) of one amino acid and the α-carboxyl (-COOH) of another amino acid. This covalent bond has a special name, called **peptide** or **amide bond**.

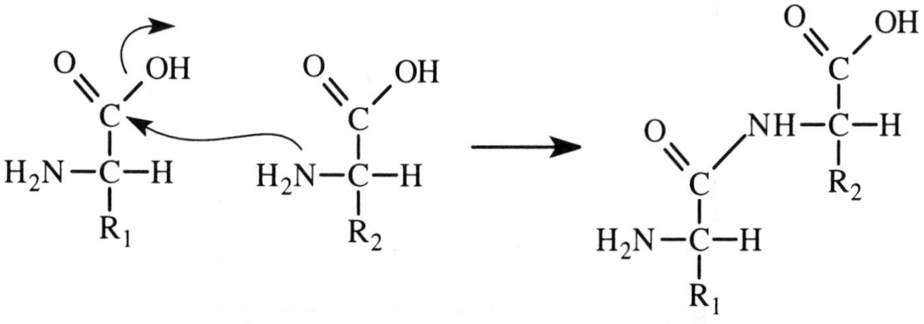

A peptide bond has some characteristics of a carbon-carbon double bond.

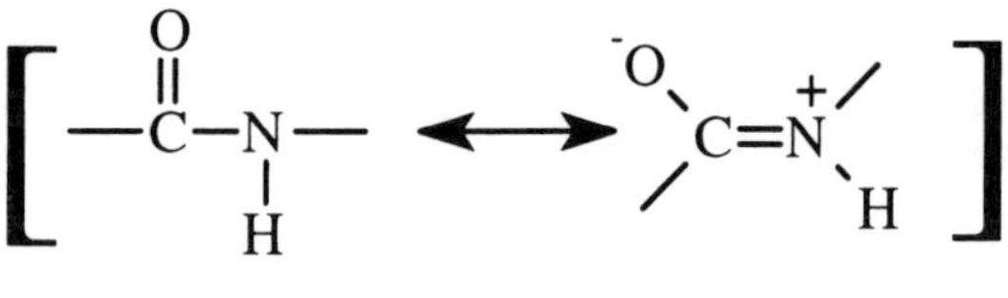

The carbon-nitrogen double bond is planar, or flat, which makes the peptide bond rigid. The planar, rigid peptide bond is an important unit for protein structure.

Peptide and protein sequences are written from left to right as amino, or N-terminus, to carboxyl, or C-terminus. The amino/carboxyl refers to free amino or free carboxyl groups, meaning that it is not part of a peptide bond.

Example:

asp-phe

H_2N-asp-phe-COOH

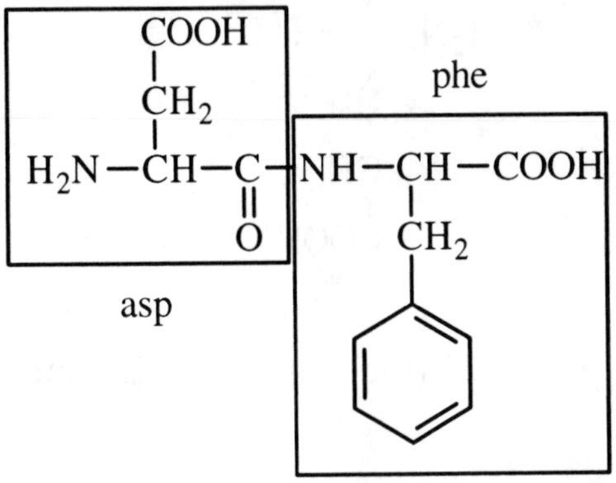

This is different from the peptide phe-asp.

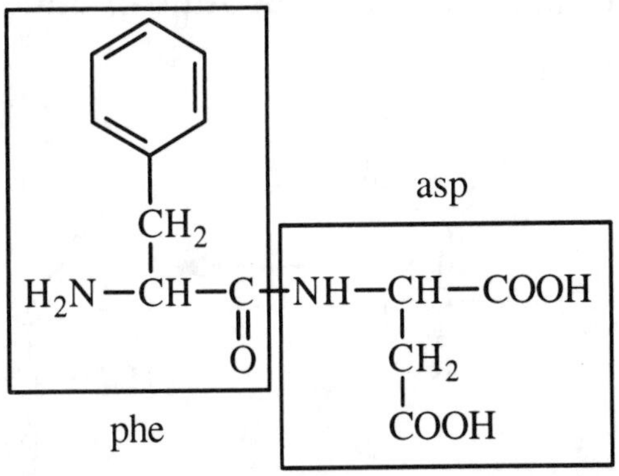

Asp-phe is an important commercial dipeptide.

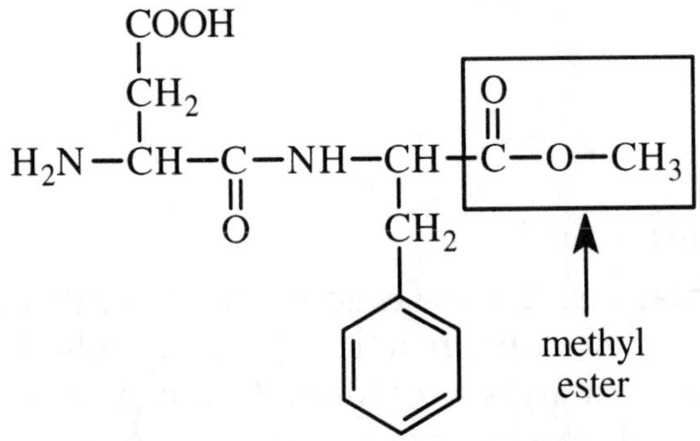

$$
\begin{array}{c}
\text{COOH} \\
| \\
\text{CH}_2 \\
| \\
\text{H}_2\text{N}-\text{CH}-\overset{\displaystyle}{\underset{\displaystyle O}{C}}-\text{NH}-\text{CH}-\overset{\displaystyle O}{\underset{}{C}}-\text{O}-\text{CH}_3 \\
\end{array}
$$

methyl ester

Aspartame™ is an artificial sweetener 200 times sweeter than sugar.

Both amino acids have the L-configuration around the α-carbon. If either is in the D-configuration, then the peptide is bitter rather than sweet. This illustrates the importance of **stereoconfiguration**.

6   Enzymes

Objectives

1. How do enzymes work?

   Enzymes are catalysts. Catalysts accelerate the rate of reaction because they provide an alternative reaction pathway which requires less energy than the uncatalyzed reaction. Catalysts work by providing a surface upon which reactants are absorbed. As reactant molecules bind to the catalyst's surface, they are oriented in a manner which increases the likelihood of product formation. As products are formed, they leave their absorption sites, which are then available for other reactant molecules.

2. How do enzymes differ from inorganic catalysts?

   The difference between inorganic catalysts and enzymes is directly related to their structure. In contrast to inorganic catalysts, each type of enzyme molecule contains a unique, intricately shaped binding surface called an **active site**. Reactant molecules, called **substrates**, bind to the enzyme's active site, which is typically a small cleft or crevice of an otherwise large protein molecule.

3. How does protein structure affect enzyme activity?

   The "lock-and-key model" assumes that the enzyme's active site has a rigid structure that is complementary in structure to the substrate. In the "induced-fit model," the structure of proteins is taken into account. In this model, substrate does not fit precisely into a rigid active site. Instead, non-covalent interactions between the enzyme and substrate cause a change in the three-dimensional structure of the active site. As a result of these interactions, the shape of the active site conforms to the shape of the substrate.

4. How are enzymes named and classified?

   Each enzyme is classified and named according to the type of chemical reaction it catalyzes. The following are the six major enzyme categories:

   Oxidoreductases: Oxidoreductases catalyze various types of oxidation-reduction reactions.

   Transferases: Transferases catalyze reactions that involve the transfer of groups from one molecule to another.

   Hydrolases: Hydrolases catalyze reactions in which the cleavage of bonds is accomplished with the addition of water.

Lyases: Lyases catalyze reactions in which groups (e.g., H_2O, CO_2, and NH_3) are removed to form a double bond, or added to a double bond.

Isomerases: This is a heterogeneous group of enzymes. Isomerases catalyze several types of intermolecular rearrangements.

Ligases: Ligases catalyze bond formation between two substrate molecules.

5. What is enzyme kinetics and how does it provide information about enzyme behavior?

The quantitative study of enzyme catalysis, referred to as enzyme kinetics, is used to provide information about reaction rates. Kinetic studies are used to measure the affinity of enzymes for substrates and inhibitors, and gain insight into reaction pathways.

6. What is enzyme inhibition?

Enzyme inhibition is a reduction or elimination of the catalytic activity.

7. What catalytic mechanisms do enzymes use to increase reaction rates?

Several important factors contribute to enzyme catalysis. They are:

Proximity and strain effects: For a biochemical reaction to occur, the substrate must come into close proximity to catalytic functional groups within the active site. Once the substrate is correctly positioned, a change in the enzyme's conformation may result in a strained enzyme-substrate complex. This strain helps to move the enzyme-substrate complex into the transition state.

Electrostatic effects: The strength of electrostatic interactions is related to the ability of surrounding solvent molecules to reduce the attractive forces between chemical groups.

Acid-base catalysis: Chemical groups can often be made more reactive by the addition or removal of a proton.

Covalent catalysis: In some enzymes, a nucleophilic side chain group forms an unstable covalent bond with the substrate. The enzyme-substrate complex then undergoes further reaction to form the product.

8. What role do metals and small organic groups play in biochemical reactions?

To catalyze reactions other than proton transfer and nucleophilic substitutions, enzymes require participation of non-protein cofactors such as metal cations and coenzymes. Because their directed valences allow them to interact with two or more ligands, metal ions facilitate the proper orientation of substrate within the active site. As a consequence of these features, the formation of the substrate-metal ion complex results in a polarization of the substrate which promotes catalysis. In addition, certain metal cations mediate redox reaction. Coenzymes, derived from vitamins, are small organic molecules that facilitate enzyme-catalyzed reactions.

9. What are allosteric enzymes?

An allosteric enzyme is one in which binding of an effector or substrate at one specific site is influenced by the binding at a different site on the protein.

10. How and why are enzyme activities regulated in the living cell?

The thousands of enzyme-catalyzed chemical reactions which occur in living cells are organized into a series of biochemical pathways. Regulation is essential for the following: maintenance of an ordered state, conservation of energy, and responsiveness to environmental changes. Regulation of biochemical pathways is achieved primarily by adjustments in the concentrations and activities of certain enzymes. Control is accomplished by utilizing various combinations of the following mechanisms: genetic control, covalent modification, allosteric regulation, and compartmentalization.

General Principles

Enzymes are biological catalysts which are capable of accelerating the rate of chemical reactions. Enzymes do not alter the thermodynamics of a reaction. In other words, if a chemical reaction is not thermodynamically favorable ($+ \Delta G$), an enzyme cannot change this. Enzymes are like the flood gates on a dam: they can control how much or how fast water flows through, but they cannot make the water flow backwards.

Enzymes accelerate the rate of reaction by stabilizing a transition state, which is the chemical intermediate of a reaction. This stabilization lowers the activation energy, which is the energy a substrate must attain before it can be transformed into product. For example, if you were popping corn in an open pot, the amount of popcorn that gets out of the pot will depend on the height of the pot. The taller the pot (or the greater the activation energy), the fewer the number of escaped kernels. The height achieved when the corn pops is not uniform because some kernels will pop higher than others. The kernels taken as a whole will have some average popping height. This would correspond to the average energy of the substrate. By lowering the height of the pot, (or, in enzyme terms, stabilizing the transition state), the proportion of the popcorn which can escape from the pot will be greater.

Enzymes can catalyze a wide variety of chemical reactions. The following are examples of general type of reactions.

Type of reaction: *Oxidation-reduction*
Enzyme class: *Oxidoreductase*
    Example: Alcohol dehydrogenase
    Function: Catalyzes the transfer of electrons from an alcohol to NAD^+, producing the corresponding aldehyde. This is the first step in the catabolism of alcohols. Electrons are removed from the alcohol carbon to form the carbonyl aldehyde.

$$CH_3\text{-}CH_2\text{-}OH + NAD^+ \rightleftharpoons CH_3-\overset{\overset{\displaystyle O}{\|}}{C}-H + NADH + H^+$$

       Ethanol   Acetaldehyde

Type of reaction: *Transfer of functional groups*
Enzyme class: *Transferase*
    Example: Catechol N-methyltransferase
    Function: Catalyzes the transfer of a methyl group from S-adenosylmethionine to norepinephrine. Used in the synthesis of the neurotransmitter epinephrine.

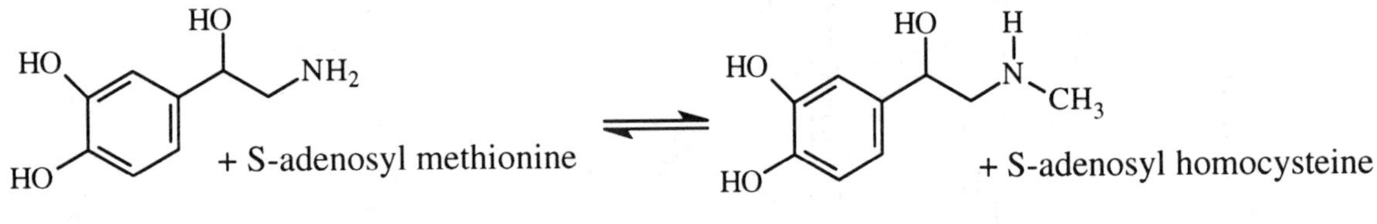

      Norepinephrine   Epinephrine

Type of reaction: *Hydrolysis, bond cleavage using water*
Enzyme class: *Hydrolyase*
    Example: Thrombin
    Function: Hydrolyzes the peptide bond after an arginine. Converts fibrinogen into fibrin; fibrin then polymerizes to form a blood clot.

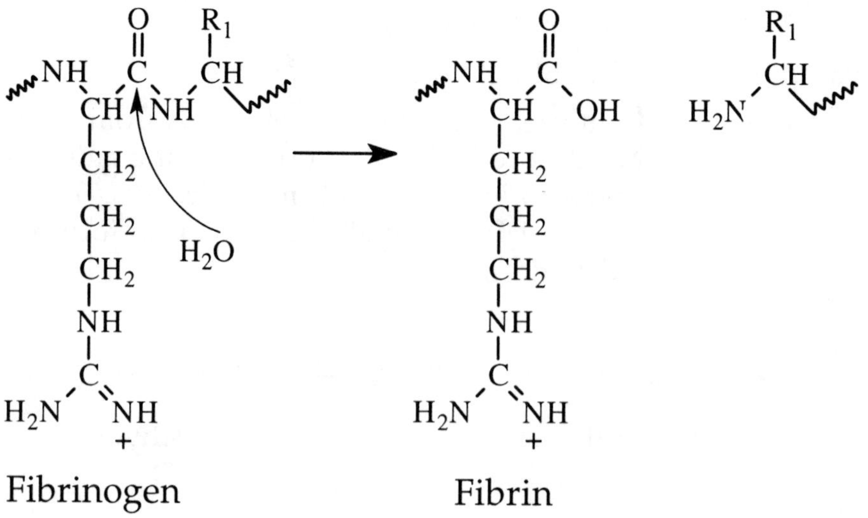

Fibrinogen   Fibrin

Type of reaction: *Group removal and formation of a double bond*

Enzyme class: *Lyase*

     Example: Argininosuccinate lyase

     Function: Converts argininosuccinate to arginine and fumarate in urea synthesis. Four carbon dicarboxylic acid is the group which is removed, forming a double bond between the central two carbons.

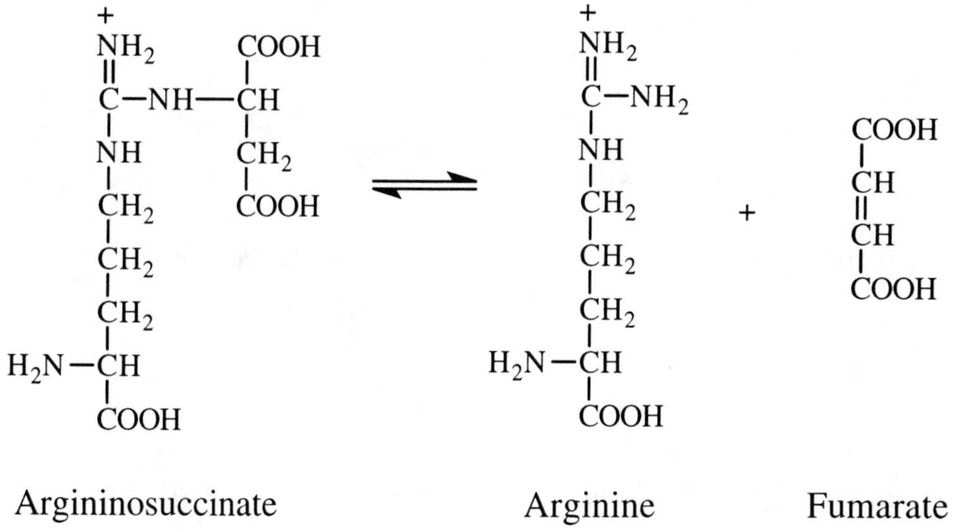

Argininosuccinate   Arginine   Fumarate

Type of reaction: *Isomerization*

Enzyme class: *Isomerase*

   Example: Triose phosphate isomerase

   Function: Interconverts glyceraldehyde-3-phosphate and dihydroxy-acetone phosphate. The carbonyl group is moved from carbon 1 in glyceraldehyde to carbon 2 in dihydroxyacetone. Involved in glycolysis or sugar fermentation.

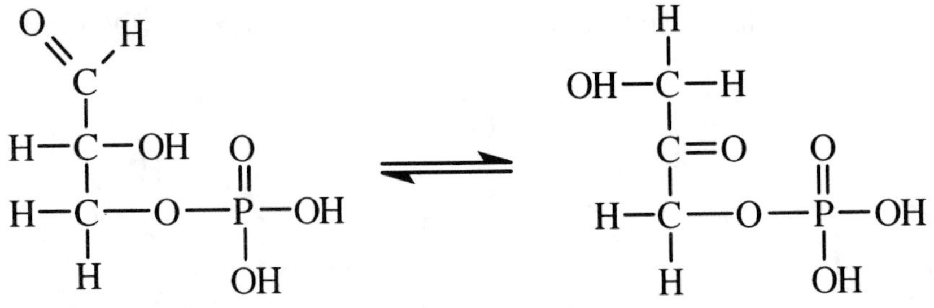

Glyceraldehyde-3-phosphate   Dihydroxyacetone phosphate

Type of reaction: *Bond formation with ATP hydrolysis*

Enzyme class: *Ligase*

   Example: Glutamine synthetase

   Function: Catalyzes the formation of glutamine from glutamate and ammonia using ATP hydrolysis to make the reaction thermodynamically favorable.

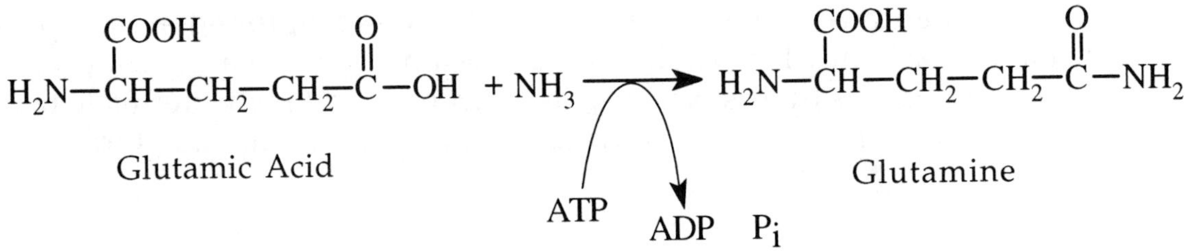

Glutamic Acid   Glutamine

ATP   ADP P_i

Vitamins and Cofactors

Enzymes can catalyze a wide variety of chemical reactions. Since enzymes are composed of twenty amino acids and essentially only the polar amino acids are chemically reactive, there is a need for additional chemical functional groups. To increase the diversity of the types of reactions catalyzed by enzymes, extra chemical groups called cofactors are often associated with enzymes.

Many vitamins serve as precursors to these cofactors. NAD$^+$ (nicotinamide adenine dinucleotide) is a cofactor which functions as an intracellular carrier of reducing equivalents (i.e., electrons). The cofactor NAD$^+$ is made from the vitamin niacin.

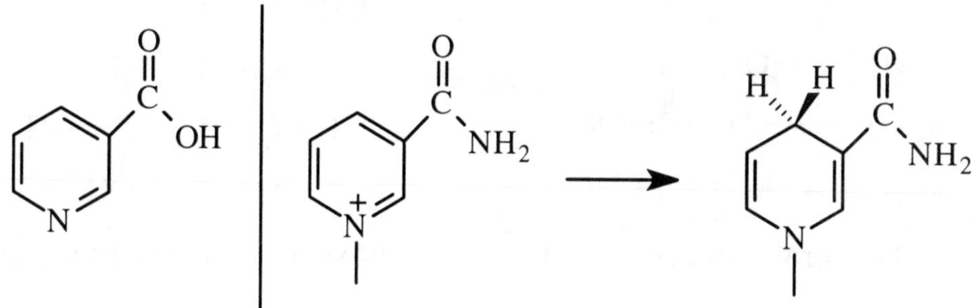

Niacin	Oxidized	Reduced
	NAD$^+$	NADH

A diet deficient in the vitamin niacin results in the disease Pellagra, characterized by diarrhea, dermatitis, and dementia. Pellagra was a significant health problem around the turn of the century in the rural south where corn was a staple food. Humans normally synthesize niacin from the amino acid tryptophan; corn, however, has very little tryptophan or usable niacin. While corn does have niacin, it is in a form that cannot be absorbed by the intestines. Native Americans, who originally domesticated corn, found a way to convert niacin into a form which could then be absorbed by humans. They accomplished this by first soaking the corn meal in lime water (calcium hydroxide), a process which converts the niacin into a useable form.

Notable differences between enzymes and chemical catalysts:

1. Enzymes are capable of higher reaction rates
2. Enzymes work under mild reaction conditions
3. Enzymes have greater reaction specificity
4. Enzymes have the capacity for regulation

Substrate specificity

The binding pocket, or active site, of an enzyme provides the specificity for determining what chemical reactions will be catalyzed by that enzyme. The substrate binds to the enzyme by non-covalent interactions. These interactions include hydrogen bonding, electrostatic, and hydrophobic forces.

Stereospecificity

Contact points on the enzyme surface have a particular orientation, similar to the way in which our right hand fits into a right-handed glove, but not a left-handed one.

Geometric specificity

Contact points within an active site, or binding pocket, have a distinctive shape. In the same way, because people have different sized feet, they wear different sized shoes. A size 12 foot, for example, cannot fit into a size 7 shoe. Likewise, because our feet differ in shape from our hands, they can not fit into gloves.

Enzyme Kinetics

In any chemical reaction, the velocity, or the change in concentration of substrate or product as a function of time, is proportional to some power of the substrate concentration.

$$v = k \times [S]^n, \text{ where } n \geq 1$$

The value of n is the order of reaction and refers to the number of reactants (or substrates) the velocity depends on. For example, if A $\rightarrow$ B and n = 1, then the reaction is referred to as first order, meaning that the velocity is dependent only on the concentration of A.

If A + B $\rightarrow$ C and n = 2, then the reaction is second order, meaning the velocity depends on the concentration of both A and B. It is important to keep in mind that the order of reaction is independent of stoichometry.

For a simple first order reaction, $v = k \times [S]$. As [S] increases, V should also increase linearly. However, enzymes are biomolecules with a finite number of binding sites. When these sites are completely filled (saturated), the enzyme cannot operate any faster. This is called the **maximal velocity** (V_{max}).

In order to study the kinetic properties of an enzyme catalyzed reaction, it is advantageous to have a mathematical equation which describes the change in velocity as a function of substrate concentration. The Michaelis-Menten equation does just that. Several assumptions were made in the initial derivation of the Michaelis-Menten equation. First, the enzyme-substrate complex (ES) is in steady state, which means the concentration of ES remains constant as a function of time. Another way to understand this is that the rate of ES formation equals the rate of its removal. The second assumption is that at saturation, all the enzyme is in the ES form (no free enzyme when the [S] >>> [E]). Third, when all the enzyme is in the ES form, the rate of product formation is maximal, $V_{max} = k_3$ [ES].

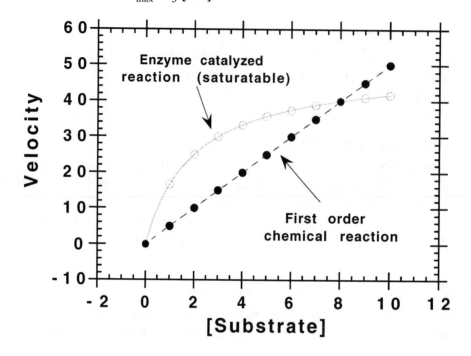

Significance of K_m and V_{max}

When [S] is equal to the value of K_m, the velocity will equal $0.5 \times V_{max}$. If the K_m value for an enzyme decreases, the concentration of substrate necessary for the velocity to equal $0.5 \times V_{max}$ decreases. One way to think about this is that it takes less substrate to generate product at a given rate.

The V_{max} is proportional to the concentration of enzyme. Remember from the second and third assumptions used to derive the Michaelis-Menten equation that at saturation all the enzyme is in the ES form, and the velocity is proportional to the [ES].

Determination of K_m and V_{max}

In order to determine the K_m, the V_{max} must be known, since K_m is the concentration of substrate at which the velocity is $0.5 \times V_{max}$. V_{max}, however, is an asymptotic value, and so the velocity can never reach V_{max}. To solve this dilemma, Lineweaver-Burke manipulated the Michaelis-Menten equation to conform to an equation with the standard form for a straight line: $y = mx + b$, where $y = \dfrac{1}{V}$, $m = \dfrac{K_m}{V_{max}}$, $x = \dfrac{1}{[S]}$, and $b = -\dfrac{1}{V_{max}}$. By plotting $\dfrac{1}{[S]}$ versus $\dfrac{1}{V}$, the x-intercept is equal to $-\dfrac{1}{K_m}$ and the y-intercept is $\dfrac{1}{V_{max}}$.

Inhibitors

Inhibitors are compounds which bind to an enzyme and interfere with its activity. Endogenously (within the cell), inhibitors are important in controlling enzyme activity. Many useful drugs are also enzyme inhibitors. For example, Lovastatin is used to treat hypercholesterolemia (high cholesterol levels) by blocking the synthesis of cholesterol. It accomplishes this by inhibiting a key enzyme in cholesterol synthesis, HMG-CoA reductase. The antibiotic penicillin is also an enzyme inhibitor. Penicillin blocks the enzyme that bacteria use to make their cell walls. Since the cell wall is important in protecting the bacteria from environmental changes, preventing its synthesis is detrimental to the bacteria. Aspirin (salicylic acid) is also an enzyme inhibitor. It prevents arachidonic acid, a membrane lipid, from being converted to prostaglandins and thromboxanes by the enzyme cyclooxygenase.

An inhibitor can work using one of two main mechanisms:

1. Competitive
2. Noncompetitive

Competitive inhibitors bind to the free enzyme; they "compete" or prevent the enzyme from binding substrate. This results in a change in K_m, but not V_{max}.

Noncompetitive inhibitors bind to a site other than the substrate binding site; they can remove both free enzyme (E) and E·S. In the simplest case, noncompetitive inhibitors will alter the V_{max} but not the K_m. This is called pure noncompetitive inhibition. Noncompetitive inhibition can not be reversed by increasing the concentration of substrate.

A third type of inhibition (not mentioned in the text) is uncompetitive. Uncompetitive inhibitors bind to E·S rather than free enzyme; they prevent the enzyme from converting E·S to product. These inhibitors will change both the K_m and V_{max}. The V_{max} decreases because the inhibitor prevents catalytic conversion. But why does the K_m decrease? One suggestion: If the inhibitor binds to E$\cdot$S and removes E$\cdot$S, then LeChatelier's principle says that the equilibrium will compensate for the change in concentration. So by removing E$\cdot$S, the inhibitor removes more substrate and enzyme, thereby decreasing the K_m.

Application of Enzyme Kinetics

Observation:

Some human populations, such as Native Americans and Asians, may be more sensitive to alcoholic beverages than others. The consumption of small amounts of ethanol in some individuals produces vasodilatation, which results in facial flushing or turning red. This physiological response to alcohol arises from acetaldehyde, which is generated by liver alcohol dehydrogenase.

$$CH_3CH_2OH + NAD^+ \longrightarrow CH_3CHO + H^+ + NADH$$

The acetaldehyde that forms is removed by a mitochondrial aldehyde dehydrogenase that converts CH_3CHO into acetate. This enzyme has a low K_m. There also exists a cytosolic aldehyde dehydrogenase with a much higher K_m.

Question:

Provide an explanation as to why certain populations may be more sensitive to alcohol than others. How would you test your explanation, and what results would you expect?

Explanation:

Individuals who are hypersensitive to ethanol often lack the mitochondrial form of aldehyde dehydrogenase. As a result, only the low affinity (high K_m) cytosolic enzyme is left to remove acetaldehyde. The concentration of acetaldehyde is thus elevated after alcohol consumption, accounting for increased sensitivity.

To test this explanation, an enzyme assay would be performed in order to measure acetaldehyde dehydrogenase activity in either the cytosol or mitochondria. One would expect to find both the high and low K_m forms of the enzyme in some individuals, and only the high K_m form in individuals sensitive to alcohol.

7 Carbohydrates

Objectives

1. How are carbohydrates classified?

   Carbohydrates are classified as monosaccharides, disaccharides, oligosaccharides, and polysaccharides on the basis of the number of simple sugar units they contain.

2. What are optical isomers and why are they important in living organisms?

   Optical isomers possess the ability to rotate plane-polarized light in opposite directions. Many biomolecules are optically active, and the ability of the enzymes to distinguish between D and L substrate molecules is an important feature of the chemistry of living cells. For example, most enzymes that are responsible for the breakdown and utilization of dietary carbohydrates can bind to D-sugars but not to their L-isomer.

3. What information do Fischer projection, Haworth, conformational, and space-filling models of carbohydrates provide?

   Fischer, Haworth, conformational, and space-filling models are structural representations of carbohydrates. The Fischer projection model shows the structure of carbohydrates as optical isomers. Haworth models represent a more accurate picture of carbohydrate ring structures with a detailed depiction of proper bond lengths and bond angles. Conformational models are more accurate than a Haworth model in that they illustrate the puckered nature of rings. Space-filling models, whose dimensions are proportional to the radius of the various atoms, also give useful structural information by indicating the volume displaced by the molecule.

4. What are the most common disaccharides found in nature and what purpose do they serve?

   Disaccharides are glycosides that are composed of two monosaccharide units. Found in abundance in nature, disaccharides provide a significant source of calories in many human diets. Examples of important disaccharides include maltose, lactose, and sucrose.

   Lactose is a disaccharide found in milk. Maltose, also known as malt sugar, is an intermediate product of starch hydrolysis and does not appear to exist freely in nature. Sucrose, common table sugar, is produced in the leaves of plants and is found throughout the plant.

5. What are the general structural features of oligosaccharides and polysaccharides?

Oligosaccharides are carbohydrates which consist of between two and ten monosaccharides units. These small polymers are most often attached to polypeptides in glycoproteins and glycolipids. Polysaccharides are composed of large numbers of monosaccharide units connected by glycosidic linkages. The most commonly occurring polysaccharides are large molecules containing hundreds to thousands of sugar units. These molecules may have a linear structure like that of cellulose or amylose, or they may have branched shapes like those found in glycogen and amylopectin. Unlike nucleic acids and proteins, which have specific molecular weights, the molecular weights of many polysaccharides do not have fixed values.

6. What structural properties of cellulose account for its widespread occurrence in plants?

Cellulose is a polymer composed of D-glucopyranose residue linked by β(1,4) glycosidic bonds. It is the most important structural polysaccharide of plants. Pairs of unbranched cellulose molecules, which may contain as many as 12,000 glucose units each, are held together by hydrogen bonding to form sheet-like strips called microfibrils. Each bundle of microfibrils contains approximately forty of these pairs.

7. What are the most common homopolysaccharides and heteropolysaccharides and what are their functions?

Homopolysaccharides are composed of only one type of monosaccharide while the heteropolysaccharides contain two or more types of monosaccharides. The homopolysaccharides that are found in abundance in nature are starch, glycogen, and cellulose. Starch, the energy reservoir of plant cells, is a significant source of carbohydrate in the human diet. Glycogen is the carbohydrate storage molecule in vertebrates. Cellulose is found in the primary and secondary cell walls of plants, where it provides a structural framework that both protects and supports cells.

Heteropolysaccharides are classified on the basis of the identity of the sugar residues they contain, the type of linkages between these residues, and the presence and location of sulfate groups. The major classifications are: hyaluronic acid, chondroitin sulfate, dermatan sulfate, heparin and heparin sulfate, and keratan sulfate. Hyaluronic acid is present in the vitreous humor of the eye and the synovial fluid of the joints. Chondroitin sulfate is an important component in cartilage. Dermatan sulfate increases during the aging process. Heparin plays a role in anticoagulant activity. Keratan sulfate is found in the cornea, cartilage, and intervertebral disks.

8.   What structural properties distinguish proteoglycans from glycoproteins?

Proteoglycans and glycoproteins belong to a group of compounds which result from the covalent linkages of carbohydrate molecules to both proteins and lipids. These compounds are collectively known as the glycoconjugates. Glycoconjugates have a profound effect on the function of individual cells, as well as the cell-cell interactions of multicellular organisms. Although both proteoglycans and glycoproteins contain carbohydrate and protein, their structures and functions appear to be substantially different.

Proteoglycans are distinguished from the more common glycoproteins by their extremely high carbohydrate content, which may constitute as much as 95% of the dry weight of such molecules. Proteoglycans aid in tissue support and elasticity in those tissues in which they occur. Consider, for example, the strength, flexibility, and resilience of cartilage. The structural diversity of proteoglycans allows them to serve a variety of roles in living organisms.

Glycoproteins are commonly defined as proteins which are covalently linked to carbohydrates. The carbohydrate composition of glycoproteins varies from 1% to over 85% of their total weight. Although glycoproteins sometimes include proteoglycans, there appears to be sufficient structural differences which necessitate examining them separately. These include the relative absence of glycoproteins in uronic acids, sulfate groups, and the disaccharide repeating units which are typical of proteoglycans.

9.   What are the functions of proteoglycans and glycoproteins?

Proteoglycans: Together with matrix proteins they form an organized meshwork which provides strengthened support to multicellular tissues. Proteoglycans are also present at the surface of cells, where they are directly bound to the plasma membrane. Although their function is not yet clear, suggestions have been made that proteoglycans play an important role in membrane structure and cell-cell interaction.

Glycoproteins: Glycoproteins occur in cells in both soluble and membrane-bound form, as well as in extracellular fluids. Vertebrate animals are particularly rich in glycoproteins. Examples of such substances include the metal-transport proteins transferrin and ceruloplasmin, blood-clotting factors, and many of the components of complement (proteins involved in cell destruction during immune reaction). Glycoproteins are now known to be important in complex recognition phenomena such as cell-molecule, cell-virus, and cell-cell interactions.

41

Study Tips

Because carbohydrates have many chiral centers, there are many stereoisomers. Identifying which carbons are chiral is the first step in determining the number of stereoisomers. Keep in mind that a chiral carbon has four different groups attached to it.

General rules:

Aldoses have n-2 chiral carbons

and

Ketoses have n-3 chiral carbons

Why?

Let's look at an example:

$$
\begin{array}{c}
O \diagdown_{} \diagup H \\
C \\
| \\
H-C-OH \\
| \\
HO-C-H \\
| \\
H-C-OH \\
| \\
H-C-OH \\
| \\
H-C-OH \\
| \\
H
\end{array}
$$

Glucose has six carbons, so n=6. Glucose is an aldose because of the aldehyde group at the top. Therefore, glucose should have n-2 chiral carbons or four chiral carbons. The first carbon has only three substituents attached to it,

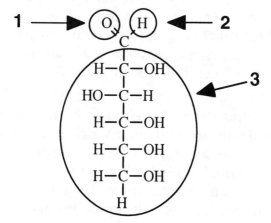

and so it is <u>not</u> chiral.

The second carbon has attached to it the following:

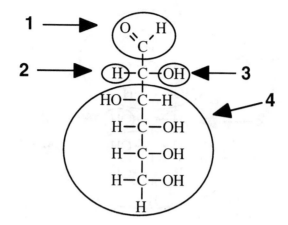

and so it is chiral.

The third carbon is as follows:

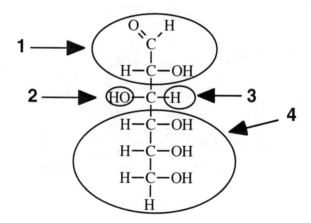

and is also chiral.

The fourth carbon,

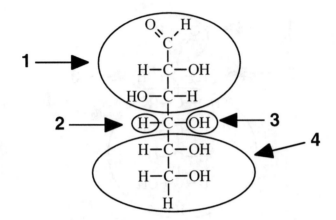

is also chiral.

The fifth carbon,

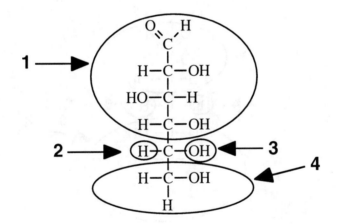

is also chiral.

The last (sixth) carbon,

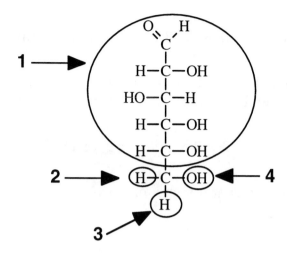

is <u>not</u> chiral.

For a carbon to be chiral it must have four different groups attached to it. The sixth carbon of glucose has two hydrogens attached to it, and so it is <u>not</u> chiral.

For glucose, therefore, there are n-2 or four chiral carbons.

Let's look at fructose (a ketose) as another example:

$$
\begin{array}{c}
H \\
| \\
H-C-OH \\
| \\
C=O \\
| \\
HO-C-H \\
| \\
H-C-OH \\
| \\
H-C-OH \\
| \\
H-C-OH \\
| \\
H
\end{array}
$$

The first carbon has two hydrogens attached to it and is therefore <u>not</u> chiral. The second carbon has only three groups attached to it and is therefore <u>not</u> chiral. The third carbon is chiral. So are the fourth and fifth carbons, but the sixth carbon has two hydrogens attached to it and is therefore <u>not</u> chiral. Thus, for fructose, carbons three, four, and five are chiral. The six carbon ketose therefore has n-3 or three chiral carbons.

The number of stereoisomers for a particular carbohydrate is related to the number of chiral carbons.

$$\text{number of isomers} = 2^x, \text{ where x=\# of chiral carbon}$$

Why is this so? Each chiral carbon can have two different arrangements of groups. For glucose, which has four chiral carbons, the number of isomers is:

$$2 \times 2 \times 2 \times 2 \text{ or } 2^4 \text{ or } 16$$

Half of these isomers are enantiomers, or mirror images, and are classified as L- or D-.

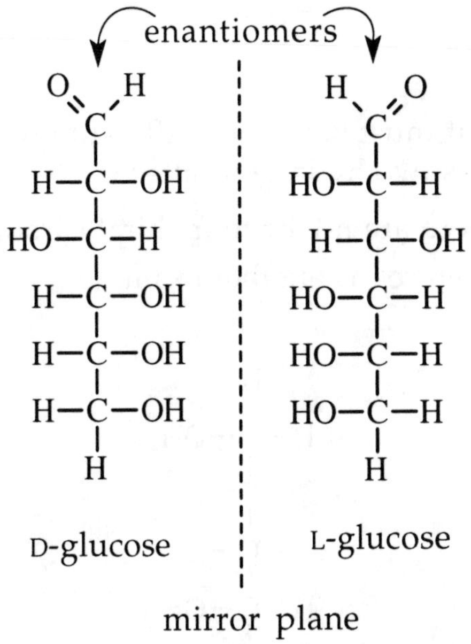

D-glucose   L-glucose

mirror plane

The remaining isomers are **diastereomers**. Diastereomers are non-enantiomeric stereoisomers, meaning they are not mirror images of each other.

A subset of diastereomers are called **epimers**. Epimers are diastereomers that differ in configuration at only one chiral center.

<u>Example:</u>

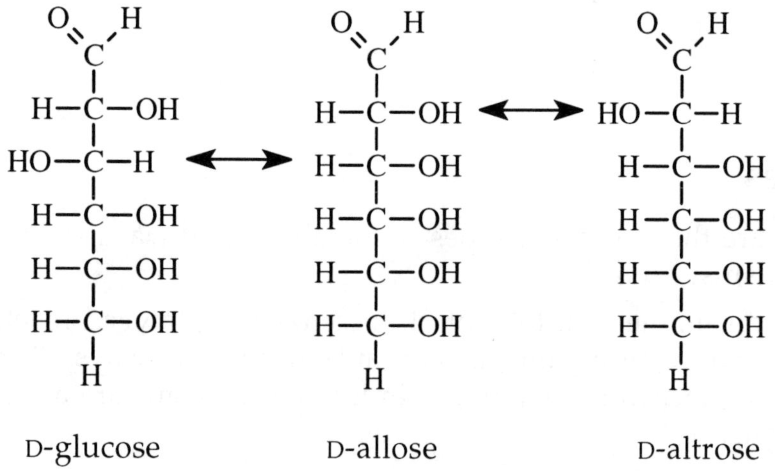

$$
\begin{array}{ccc}
\underset{C}{\overset{O\diagdown\diagup H}{\parallel}} & \underset{C}{\overset{O\diagdown\diagup H}{\parallel}} & \underset{C}{\overset{O\diagdown\diagup H}{\parallel}} \\
H-C-OH & H-C-OH \longleftrightarrow HO-C-H \\
HO-C-H \longleftrightarrow H-C-OH & H-C-OH \\
H-C-OH & H-C-OH & H-C-OH \\
H-C-OH & H-C-OH & H-C-OH \\
H-C-OH & H-C-OH & H-C-OH \\
H & H & H
\end{array}
$$

D-glucose	D-allose	D-altrose

D-glucose and D-allose are **epimers** of each other. They differ by one chiral center, or, the third carbon. Likewise, D-allose and D-altrose are epimers of each other, being different in configuration around the second carbon. However, note that D-glucose and D-altrose are <u>not</u> epimers of each other because they are different at both the second and third carbons.

8   Carbohydrate Metabolism

Objectives

1. What are the two major types of metabolic pathways in living organisms?

    Metabolism, the sum total of all the enzyme catalyzed reactions that occur in a living organism, is organized into pathways. The two major types of biochemical pathways in living organisms are anabolic and catabolic.

2. What biochemical reactions are responsible for the breakdown of food molecules in energy production?

    The biochemical pathways responsible for the breakdown of food molecules for the production of energy occur in two major pathways: glycolysis and the citric acid cycle.

    The fate of glucose depends in part on the metabolic state of the animal. Immediately after a meal when blood glucose levels are high, glycogen, a storage form of glucose, is synthesized via the glycogenesis pathway. As the levels of glucose decrease, glycogenolysis begins. Glycogenolysis ensures that there is a steady supply of glucose in the bloodstream by degrading the glycogen. Glycolysis then converts the glucose into pyruvate, resulting in the production of a small amount of energy. Glucose can also be synthesized from pyruvate by gluconeogenesis when the liver glycogen is depleted.

3. How are metabolic processes regulated so that energy and biosynthetic requirements of organisms are consistently met?

    The answer to this question is not yet fully understood. However, various forms of intercellular communication occur by means of chemical signals. Once released into the extracellular environment, each chemical signal is recognized by specific cells, which then respond in a specific manner. In animals, the nervous system and endocrine system are primarily responsible for the coordination of metabolism. The nervous system provides a rapid and efficient mechanism of acquiring and processing environmental information. Neurons release neurotransmitters and the subsequent binding of neurotransmitter molecules to nearby cells results in specific responses from those cells.

    Metabolic regulation by the endocrine system is achieved by the secretion of chemical signals, called hormones, directly into the blood. Most hormone-induced changes in cell function result from alterations in the activity or concentration of enzymes. Hormones interact by binding to specific receptor molecules.

4. How is glycogen alternately synthesized and degraded to provide animals with a consistent supply of glucose?

Both synthesis and degradation are controlled through a complex mechanism involving insulin, glucagon, and epinephrine. The binding of glucagon to liver cells stimulates glycogen degradation and inhibits glycogen synthesis. As blood glucose levels drop in the hours after a meal, glucagon ensures that glucose will be released into the bloodstream. The binding of insulin to receptors on the surface of several cell types stimulates glycogen synthesis and inhibits glycogen degradation. Epinephrine promotes glycogen degradation and inhibits glycogen synthesis. When an animal is frightened or startled, for example, epinephrine is released in relatively large quantities, resulting in an increase in blood glucose and thereby prompting the animal to either fight or take flight.

5. How do cells extract energy from glucose?

Energy is obtained from glucose via glycolysis, which consists of ten reactions.

a. Glucose is phosphorylated and cleaved to form two molecules of glyceraldehyde-3-phosphate (G-3-P). The two ATP molecules which are consumed during this stage can be considered an "investment," since the phosphoryl groups in the two G-3-P molecules will subsequently be used in ATP synthesis.

b. G-3-P is converted to pyruvate. Four ATP molecules and two NADH are produced. Because two ATP were consumed in the first series of reactions, the net production of ATP per glucose molecule is two.

6. What is the difference between aerobic respiration and fermentation?

Aerobic respiration utilizes oxygen as a terminal electron acceptor in the conversion of pyruvate into carbon dioxide and water, while fermentation regenerates NAD^+ by converting pyruvate into lactate or ethanol.

7. How do glycolysis and gluconeogenesis differ? How are they similar?

Similarities

Seven of the ten steps in glycolysis are the same in gluconeogenesis. Three irreversible reactions in glycolysis are reversed in gluconeogenesis by alternative mechanisms.

<u>Differences</u>

Gluconeogenesis converts pyruvate into glucose, while glycolysis converts glucose into pyruvate.

Insulin suppresses the synthesis of all of the key gluconeogenic enzymes and at the same time promotes glycolysis by stimulating the synthesis of fructose-2,6-bisphosphate.

8.   What regulatory mechanism ensures that glycolysis and gluconeogenesis do not occur simultaneously to any great extent?

Hormones affect gluconeogenesis and glycolysis by altering the concentrations of allosteric effectors and the rate at which key enzymes are synthesized.

Study Tips

<u>Glycolysis</u>: Focus on the structure of the intermediates and at each step note any changes which occur to the molecule. Keep in mind the function of the pathway and how each reaction brings you closer to the desired end product.

An important tip in studying metabolism is to **never** forget about the **function** of the pathway. Begin by visualizing the structure of the starting material and follow the transformations which occur until you get to the final product of the pathway. This simplifies remembering the sequence or order of reactions.

What are the functions of glycolysis?

<u>Functions:</u>

1.   Generate energy (ATP)

2.   Generate intermediates

This is illustrated by the overall reaction:

$$\text{glucose} + 2\ \text{ADP} + 2\ P_i \longrightarrow 2\ \text{lactic acid} + 2\ \text{ATP} + 2\ H_2O$$

ATP (energy) is produced by glycolysis; lactic acid (intermediates) is also produced. Note that 1-six-carbon glucose molecule is converted into 2-three-carbon lactic acid molecules.

Next, group the reactions of glycolysis into "stages," according to the type of modifications.

Stage 1 – Priming stage: readying glucose for metabolism.

This occurs in three steps and requires ATP (hint: "It takes energy to make energy.")

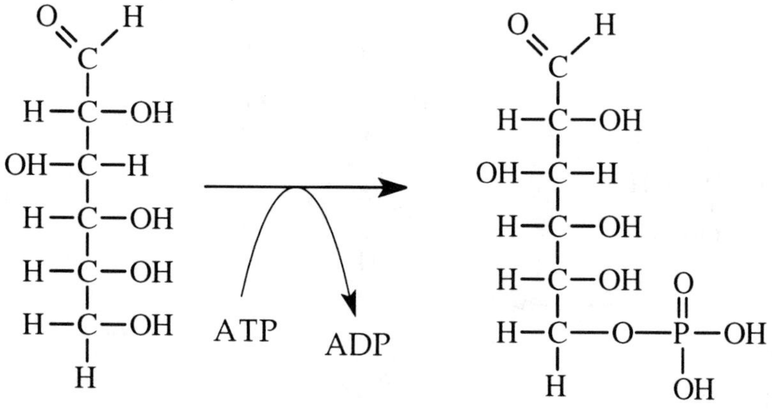

ATP → ADP

Glucose

Glucose-6-phosphate

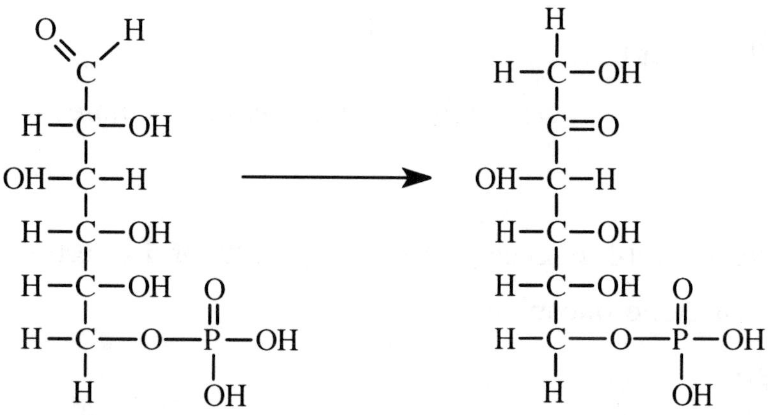

Glucose-6-phosphate → Fructose-6-phosphate

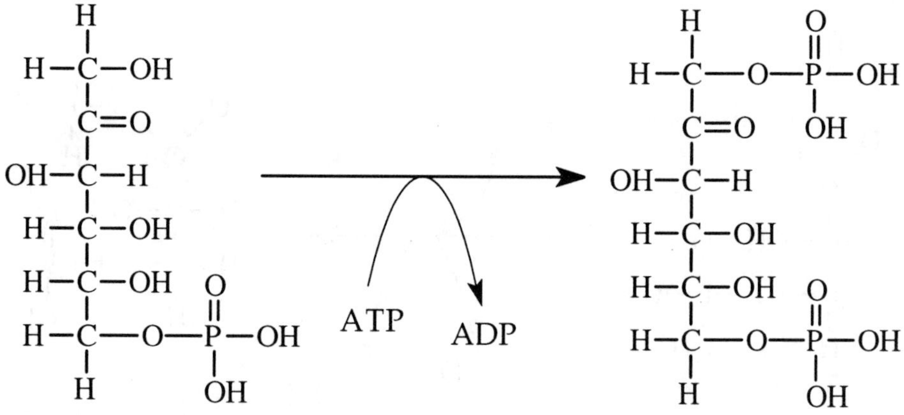

Fructose-6-phosphate

ATP → ADP

Fructose-1,6-bisphosphate

Stage 2 – Splitting stage: cutting the 1-six-carbon molecule into 2-three-carbon molecules.

Dihydroxyacetone phosphate

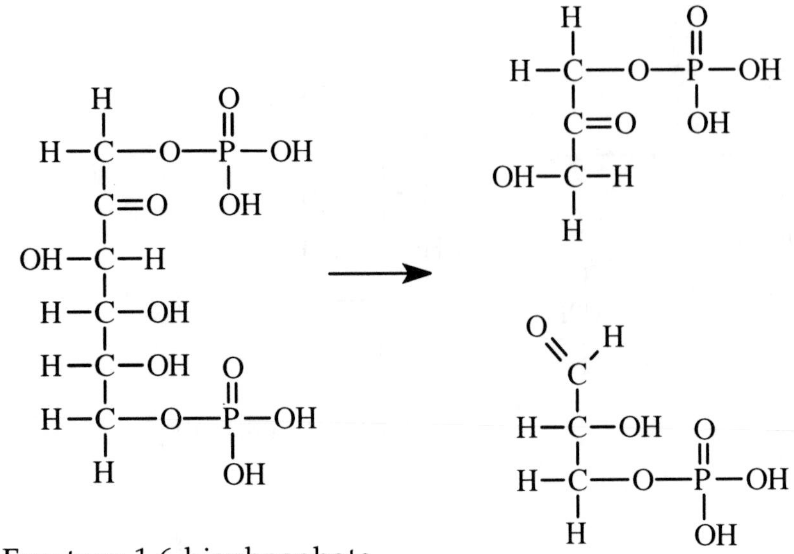

Fructose-1,6-bisphosphate

Glyceraldehyde-3-phosphate

Stage 3 – Energy recapture stage (Oxidation-Reduction-Phosphorylation)

Dihydroxyacetone phosphate

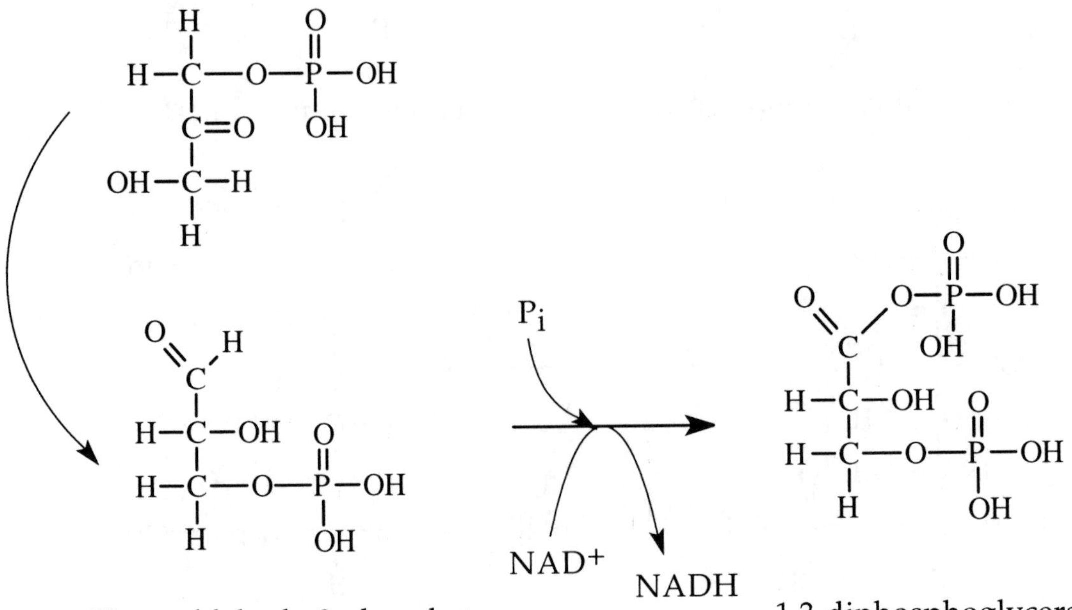

P_i

NAD^+ NADH

Glyceraldehyde-3-phosphate

1,3-diphosphoglycerate

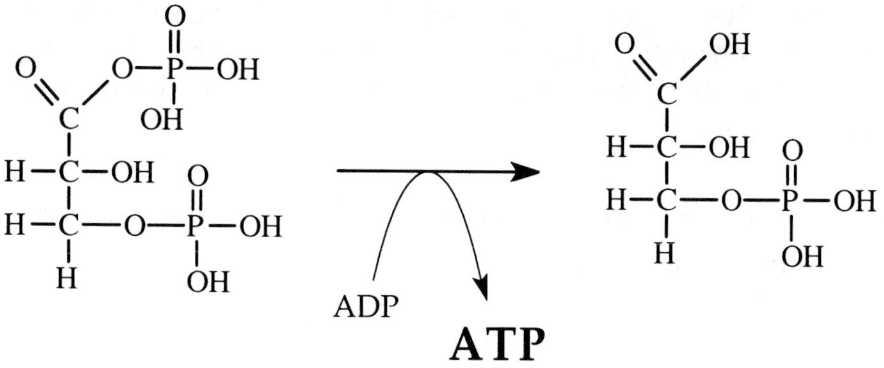

1,3-diphosphoglycerate → 3-phosphoglycerate (ADP → ATP)

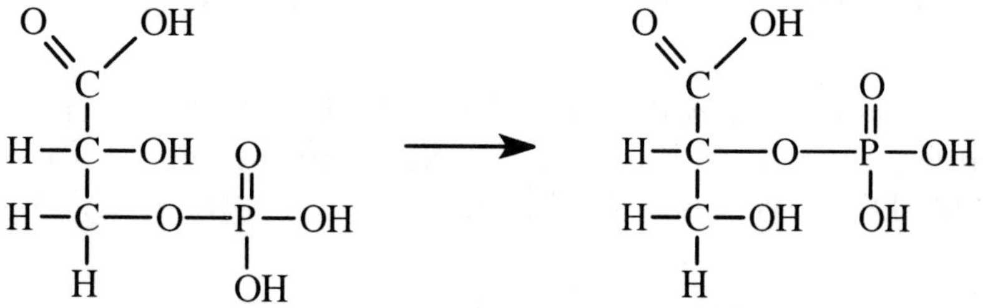

3-phosphoglycerate → 2-phosphoglycerate

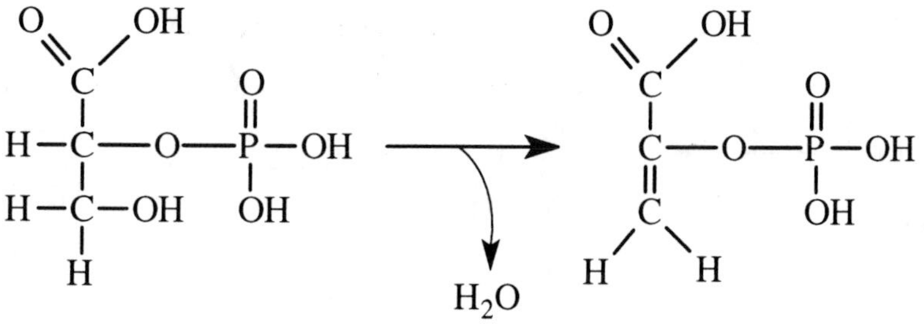

2-phosphoglycerate → phosphoenolpyruvate (H_2O)

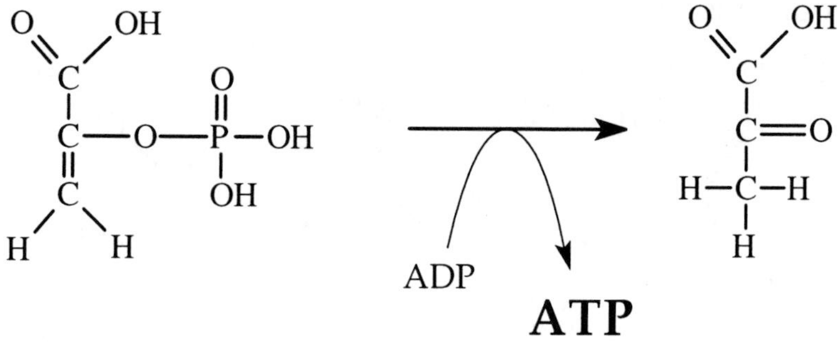

phosphoenolpyruvate → pyruvate (ADP → ATP)

Remember: glycolysis is not the only set of reactions occurring in the cell; many other reactions are happening simultaneously. The kinds of reactions occurring and the fate of the intermediates depend on the needs and requirements of the cell.

For example, pyruvate is an intermediate in glycolysis and its fate depends on:
1. whether oxygen is present
2. type of cell

If oxygen is present, pyruvate can enter the citric acid cycle to be oxidized to CO_2 (Chapter 11). Additionally, the electrons (NADH) generated by glycolysis can be transferred to O_2 by the electron transport chain (Chapter 11).

If no oxygen is present, the cell needs to recycle the electron carrier NAD^+. (Note: There is only a limited amount of NAD^+ in the cell so if it were all converted to NADH, glycolysis would cease.)

The process by which NAD^+ is recycled depends on the type of cell.

In yeast, for example,

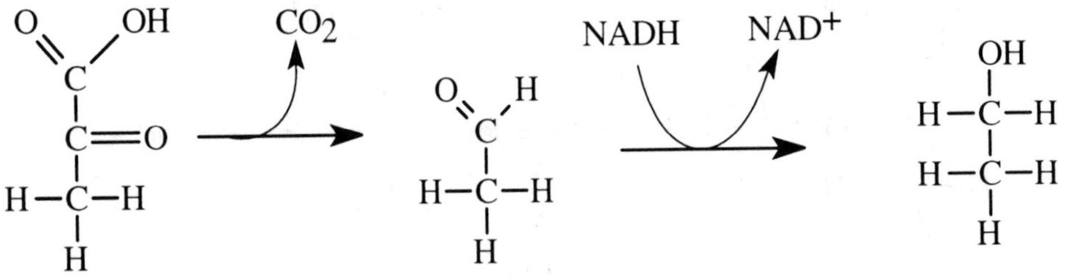

pyruvate   acetaldehyde   ethanol

Whereas, in our muscles,

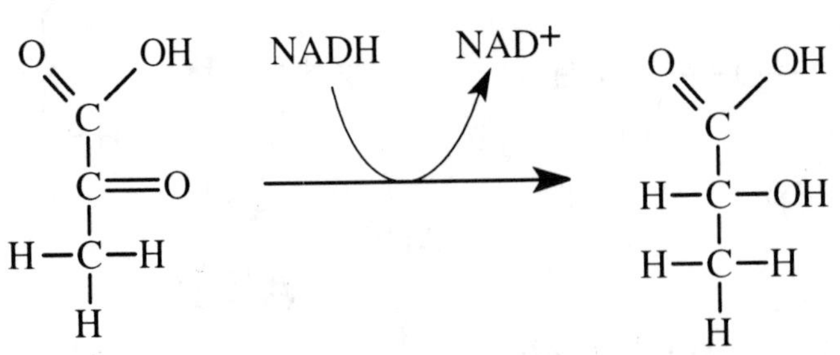

pyruvate   lactate

54

Glycogen metabolism

Glycogen (animals) – storage polymer of glucose
Starch (plants) – storage polymer of glucose

Most glycogen is found in the muscle and liver:

  ≈10% mass of liver
  ≈1% mass of muscle

The actual amount of glycogen depends on the nutritional state of the organism.

Glycogen is made of α-1 $\rightarrow$4 linked glucose, and α-1 $\rightarrow$6 linked branches

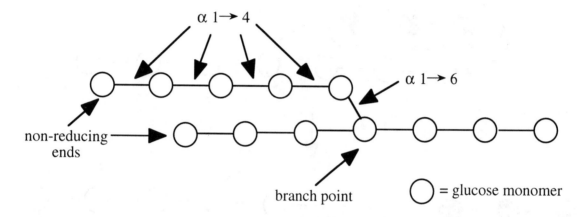

Non-reducing end: when anomeric carbon (or C-1 carbon) of the glucose is in a glycosidic linkage

Glycogenolysis: degradation of glycogen. The enzymes of glycogenolysis are similar in liver and muscle but they **function** quite differently!

  Muscle: glycogen is used to produce glucose for use in glycolysis
             and TCA for energy.
  Liver: glycogen is used to produce glucose for export out of liver
             for other cells to use.

Basic enzymatic reaction:

  Glycogen phosphorylase

$$\underset{(n\text{ residues})}{\text{glycogen}} + P_i \longrightarrow \underset{(n\text{-1 residues})}{\text{glycogen}} + \text{glucose-1-phosphate}$$

55

Glycogen phosphorylase catalyzes the progressive removal from <u>non-</u>reducing ends. The enzyme stops four units prior to a branch.

This leaves a "limit dextrin," or a small branched polymer.

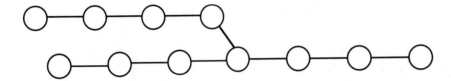

The limit dextrin is further degraded by the glycogen debranching enzymes glucanotransferase and 1,6-glucosidase.

1.   glucanotransferase

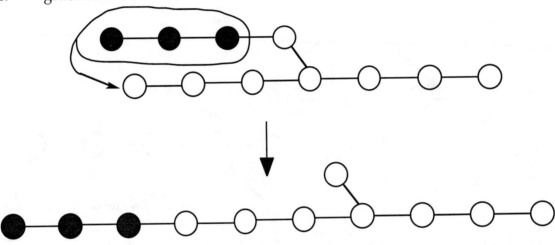

2.   1,6-glucosidase

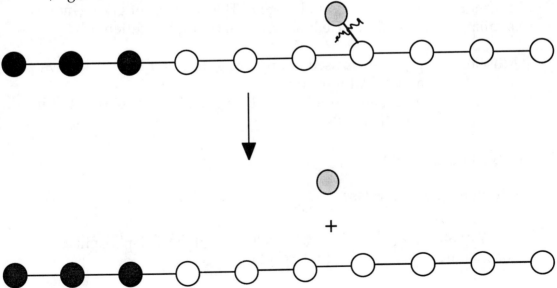

The unbranched polymer produced by 1,6-glucosidase is then degraded by glycogen phosphorylase.

The glucose-1-phosphate produced by glycogen phosphorylase is isomerized by phosphoglucomutase.

$$\text{glucose-1-phosphate} \rightleftharpoons \text{glucose-6-phosphate}$$

In muscle, glucose-6-phosphate is catabolized by glycolysis.

Since the function of glycogenolysis in liver is to produce glucose for export out of the liver, the fate of glucose-6-phosphate is as follows:

$$\text{glucose-6-phosphate} + H_2O \longrightarrow \text{glucose} + P_i$$

Glycogen synthesis is not a simple reversal of glycogen degradation.

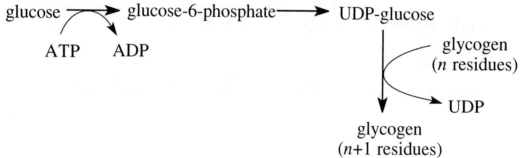

Glycogen synthase:

$$\text{UDP-glucose} + \underset{(n \text{ residues})}{\text{glycogen}} \longrightarrow \underset{(n+1 \text{ residues})}{\text{glycogen}} + \text{UDP}$$

Note: Glucose is added to the non-reducing end.

Gluconeogenesis

Gluconeogenesis is an **anabolic** pathway used to synthesize glucose from amino acids or lactate. Although at first glance it may appear that gluconeogenesis is the reverse of glycolysis, this would be thermodynamically impossible.

Most reactions of glycolysis have $\Delta G°$ close to zero. Therefore those reactions are reversible.

Three key reactions have large $\Delta G°$ and are essentially irreversible:

1.   glucose + ATP $\longrightarrow$ glucose-6-phosphate + ADP $\Delta G°= -4\ \dfrac{kcal}{mol}$

To reverse this reaction, the cell uses a different enzyme and a different chemical reaction, catalyzed by glucose-6-phosphatase.

glucose-6-phosphate + H_2O $\longrightarrow$ glucose + P_i $\Delta G°= -3.3\ \dfrac{kcal}{mol}$

2.   fructose-6-phosphate + ATP $\longrightarrow$ fructose-1,6-bisphosphate + ADP

$\Delta G°= -3.4\ \dfrac{kcal}{mol}$

The phosphofructokinase reaction diagrammed above is reversed in gluconeogenesis by the enzyme fructose-1,6-bisphosphatase:

fructose-1,6-bisphosphate + H_2O $\longrightarrow$ fructose-6-phosphate + P_i

$\Delta G°= -3.9\ \dfrac{kcal}{mol}$

3.   PEP + ADP $\longrightarrow$ pyruvate + ATP $\Delta G°= -7.5\ \dfrac{kcal}{mol}$

The pyruvate kinase reaction diagrammed above is reversed by a series of reactions. In mammals, these reactions occur in both the mitochondrion and the cytosol. The overall reaction is as follows:

pyruvate + ATP + GTP $\longrightarrow$ PEP + ADP + GDP + P_i $\Delta G°= +0.2\ \dfrac{kcal}{mol}$

Study Tips

The upper reactions are glycolysis, the lower are gluconeogenesis:

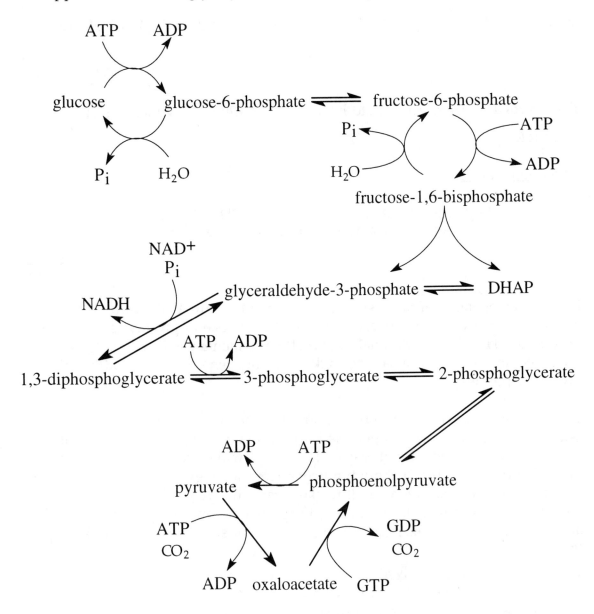

<u>Energetics of the two pathways</u>

<u>Glycolysis:</u>

glucose + 2 ADP + 2 P_i + 2 NAD^+ $\longrightarrow$ 2 pyruvate + 2 ATP + 2 NADH + 2 H_2O

$$\Delta G° = -20 \; \frac{kcal}{mol}$$

The reversal of glycolysis would have a $\Delta G°$ of +20 $\frac{kcal}{mol}$, which is <u>not</u> favorable.

<u>Gluconeogenesis:</u>

2 pyruvate + 4 ATP + 2 GTP + 2 NADH + 2 H_2O $\longrightarrow$
glucose + 4 ADP + 2 GDP + 6 P_i + 2 NAD^+

$$\Delta G° = -9 \; \frac{kcal}{mol}$$

For gluconeogenesis, it takes six ATP equivalents (four ATP + two GTP) to convert pyruvate $\longrightarrow$ glucose, whereas the reversal of glycolysis would use only two ATPs. The additional four ATP equivalents used in gluconeogenesis are necessary to make gluconeogenesis a thermodynamically favorable pathway.

In glycolysis, glucose $\longrightarrow$ pyruvate, and in gluconeogenesis, pyruvate $\longrightarrow$ glucose. However, gluconeogenesis is <u>not</u> a reversal of glycolysis! In biology, organisms have evolved pathways which are **always** thermodynamically favorable (i.e., $\Delta G°<0$). Although gluconeogenesis does use some of the same enzymes as glycolysis, there are three key steps in glycolysis which are essentially irreversible (hexokinase, phosphofructokinase, pyruvate kinase) and therefore gluconeogenesis utilizes a different set of reactions to overcome this irreversibility.

1. <u>Hexokinase (glycolysis)</u>
    Glucose + ATP $\longrightarrow$ Glucose-6-P + ADP

    Glucose-6-phosphatase (gluconeogenesis)
    Glucose-6-phosphate + H_2O $\longrightarrow$ Glucose + P_i

2. <u>Phosphofructokinase (glycolysis)</u>
    Fructose-6-P + ATP $\longrightarrow$ Fructose-1,6-bisphosphate + ADP

    Fructose-1,6-bisphosphatase (gluconeogenesis):
    Fructose-1,6-bisiphosphate + H_2O $\longrightarrow$ Fructose-6-phosphate + P_i

3. <u>Pyruvate kinase (glycolysis)</u>

    PEP + ADP $\longrightarrow$ pyruvate + ATP

    Pyruvate carboxylase/PEP carboxylkinase (gluconeogenesis):

    a) pyruvate + CO_2 + ATP + H_2O $\longrightarrow$ oxaloacetate + ADP + P_i

    b) oxaloacetate + GTP $\longrightarrow$ PEP + GDP + CO_2

    _____

    pyruvate + ATP + GTP + H_2O $\longrightarrow$ PEP + ADP + GDP + P_i

9   Lipids and Membranes

Objectives

1. What are the major lipid classes?

    The major lipid classes are: fatty acids and their derivatives, triacylglycerols, wax esters, phospholipids, sphingolipids, and isoprenoids.

2. How do the structural properties of saturated and unsaturated fatty acids affect their biological properties?

    Fatty acid chains that contain only carbon-carbon single bonds are referred to as **saturated**. Those molecules that contain one or more double bonds are said to be **unsaturated**. Because double bonds are rigid structures, molecules which contain them occur in the *cis* or *trans* isomeric forms. The double bonds in most naturally occurring fatty acids are in *cis*-configuration. Because of this structural feature, unsaturated fatty acids do not pack as closely together as saturated fatty acids and therefore have a lower melting point.

3. What are eicosanoids and why are they important in mammals?

    Eicosanoids are a diverse group of extremely powerful, hormone-like molecules which are produced in most mammalian tissues. They are generally active within the organ in which they are produced; eicosanoids are therefore called autocrine regulators instead of hormones. The three major eicosanoids are: prostaglandins, thromboxanes, and leukotrienes. Prostaglandins promote inflammation, an infection-fighting process which produces pain and fever. They are also involved in reproductive processes and digestion. Thromboxane is produced primarily by platelets (blood cells which initiate blood clotting). Once released, thromboxane promotes platelet aggregation and smooth muscle contraction. Leukotrienes attract infection-fighting white blood cells to damaged tissue.

4. Why are the triacylglycerols an efficient form of stored energy?

    Triacylglycerol molecules are a more efficient form of energy than glycogen for several reasons:

    a. Because triacylglycerols are hydrophobic, they coalesce into compact anhydrous droplets within cells. A substantial amount of water is associated with glycogen. The anhydrous triacylglycerols can store eight times as much energy as an equivalent amount of glycogen.

b. Triacylglycerol molecules are less oxidized than carbohydrate molecules. As a result, the oxidation of triacylglycerol releases more energy.

5. What role do wax esters play in living organisms?

Wax esters are complex mixtures of non-polar lipids. They serve as a protective coating on leaves, stems, and fruits of plants, and the skin and fur of animals.

6. What structural properties of phospholipids are responsible for their role as the major constituent of biological membranes?

Phospholipids have several roles in living organisms. They are first and foremost structural components of membranes. Several phospholipids also serve as emulsifying agents and surface active agents. Phospholipids are well-suited for these roles because they are amphipathic molecules. Despite their structural differences, all phospholipids have hydrophobic and hydrophilic domains. The hydrocarbon chains are the hydrophobic domain, while the hydrophilic domain is composed of phosphate and often some other polar or charged group.

7. How do sphingolipids and glycolipids differ from phospholipids? What role do these molecules play in living organisms?

Sphingolipids are important components of animal and plant membranes. All sphingolipid molecules contain a long chain amino alcohol. In animals, this is primarily sphingosine. The core structure of a sphingolipid is ceramide, which is a fatty acid amide derivative of sphingosine.

Glycolipids are sphingolipids which contain carbohydrate groups and no phosphate. An important difference between phospholipids, glycolipids, and sphingolipids is that phospholipids are ionic, while the other two lipids are nonionic.

Sphingolipids are found in the cell membranes. Sphingomyelin is found in greatest abundance in the myelin sheath of nerve cells where it facilitates rapid transmission of nerve impulses. The role of glycolipids is unclear. Certain glycolipid molecules have been implicated in the binding of bacterial toxins, as well as bacterial cells, to animal cell membranes.

8. What structural feature do terpenes and steroids have in common? How do these isoprenoid molecules differ from each other? What roles do they play in living organisms?

Isoprenoids are a vast array of biomolecules which contain repeating five-carbon structural units known as isoprene units. Isoprenoids consist of terpenes and steroids. Terpenes are an enormous group of molecules which are found largely in the essential oils of plants.

Steroids are triterpene derivatives of a complex hydrocarbon ring system.

Terpenes are classified according to the number of isoprene residues, while steroids are distinguished by double bonds in the four fused rings.

Several important biomolecules are composed of nonterpene components attached to isoprenoid groups. Examples of these biomolecules include vitamin E, vitamin K, and some plant hormones.

In addition to its role as an essential component in animal cell membranes, cholesterol is a precursor in the biosynthesis of all steroid hormones, as well as vitamin D and bile salts.

9. What role do lipids and proteins play in membrane structure and function?

The basic structure of a membrane is a bimolecular lipid layer. Proteins, most of which reside within the lipid bilayer, largely determine a membrane's biological functions.

Phospholipids form bimolecular layers when they are present in sufficient concentrations. It is this property of phospholipids which is the basis of membrane structure. Membrane lipids are largely responsible for several other important features of biological membranes such as:

a. Membrane fluidity: The term fluidity describes the resistance of membrane components to movement. Rapid lateral movement is thought to be responsible for the proper functioning of many membrane proteins.

b. Selective permeability: Because of their hydrophobic nature, the hydrocarbon chains in lipid bilayers provide a virtually impenetrable barrier to ionic and polar substances.

c. Self-sealing capabilities: When lipid bilayers are disrupted, they immediately and spontaneously reseal.

d. Asymmetry: Biological membranes are asymmetric, meaning that the lipid composition of each half of a bilayer is different.

Membrane proteins are usually classified by the type of function they perform. Some of these functions include membrane structure, catalysis, signaling, and transport.

Study Tips

Fatty acids are amphiphilic, meaning they contain both hydrophobic and hydrophilic ends. Fatty acids are comprised of a polar end (carboxylic acid) and a non-polar end (hydrocarbon chain).

Most natural fatty acids have an even number of carbons (this is discussed in Chapter 10).

If the hydrocarbon chain can have double bonds within it, the fatty acid is considered **unsaturated** (not completely filled with hydrogens). If the hydrocarbon chain contains no double bonds, it is referred to as **saturated** (completely filled with hydrogens).

The number of carbons in the hydrocarbon chain affects the melting point of the fatty acid (i.e., whether the fatty acid is a solid or liquid at a particular temperature). The longer the acyl chain, the higher the melting point or the more "solid-like" is the fatty acid.

In saturated fatty acids, for example:

myristic acid	C-14	$CH_3(CH_2)_{12}COOH$	mp = 52 °C
palmitic acid	C-16	$CH_3(CH_2)_{14}COOH$	mp = 63 °C

The degree of unsaturation (number of double bonds in the hydrocarbon chain) will also affect the fluidity (melting point) of the fatty acid. The greater the number of double bonds, the lower the melting point.

Unsaturation of the eighteen carbon fatty acids, for example, have the following properties:

oleic acid
   1 double bond $CH_3(CH_2)_7CH=CH(CH2)_7COOH$ mp = 13°C
linoleic acid
   2 double bonds $CH_3(CH_2)_4CH=CHCH_2CH=CH(CH2)_7COOH$ mp = -9°C

Fatty acids are attached to glycerol via an ester bond.

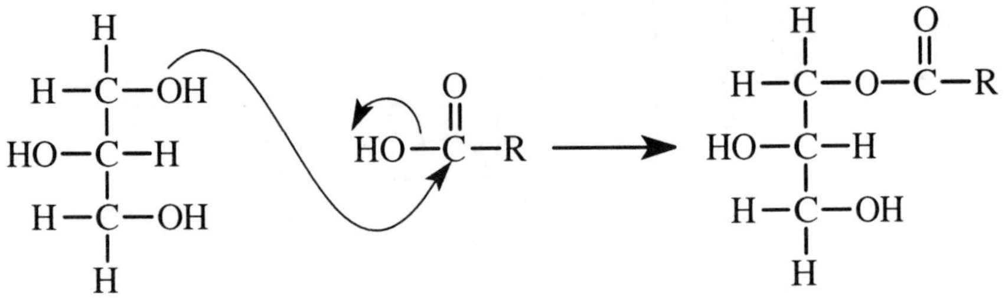

glycerol   fatty acid

Each hydroxyl group of glycerol can have a fatty acid attached to it. When three fatty acids are bonded to glycerol, the lipid is called **triacylglycerol**. Triacylglycerols are important in fat storage.

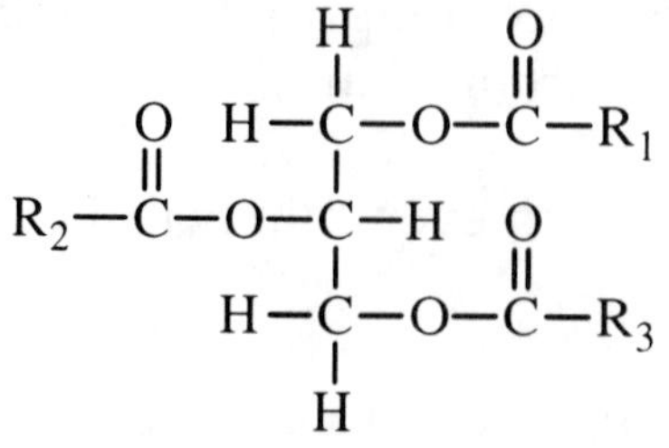

$$
\begin{array}{c}
\underset{\displaystyle H}{|} \qquad \overset{\displaystyle O}{\|} \\
O \quad H-C-O-C-R_1 \\
\| \qquad | \\
R_2-C-O-C-H \quad O \\
| \qquad \| \\
H-C-O-C-R_3 \\
| \\
H
\end{array}
$$

Another important family of fatty acyl glycerols is the phosphoacyl glycerides or **phospholipids**. Phospholipids are comprised of a glycerol "backbone," two fatty acyl chains, and a phosphate, and are an important component of biological membranes.

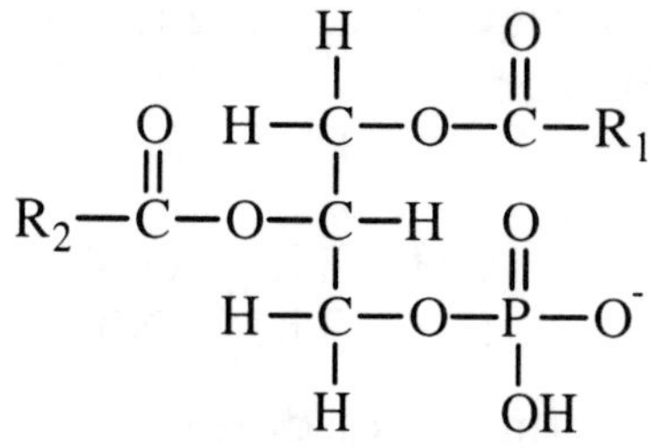

$$
\begin{array}{c}
H \qquad O \\
| \qquad \| \\
O \quad H-C-O-C-R_1 \\
\| \qquad | \\
R_2-C-O-C-H \quad O \\
| \qquad \| \\
H-C-O-P-O^- \\
| \qquad | \\
H \qquad OH
\end{array}
$$

Phosphatidic acid

Other types of molecules can be attached to the phosphate group via a phosphodiester bond linkage. The molecules are sometimes referred to as **head groups**.

Example:

phosphatidyl ethanolamine

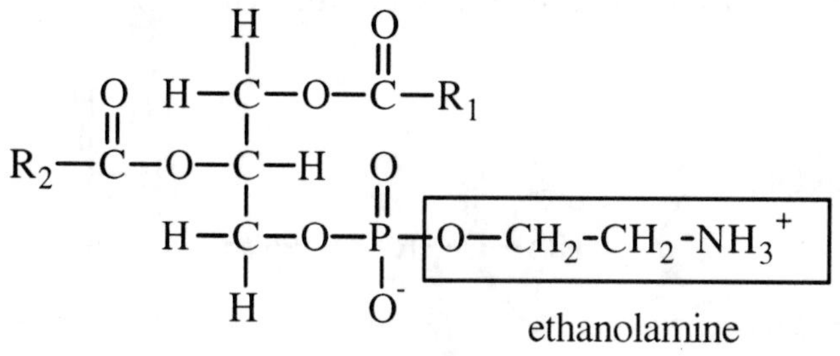

$$
\begin{array}{c}
H \qquad O \\
| \qquad \| \\
O \quad H-C-O-C-R_1 \\
\| \qquad | \\
R_2-C-O-C-H \quad O \\
| \qquad \| \\
H-C-O-P-\boxed{O-CH_2-CH_2-NH_3^+} \\
| \qquad | \\
H \qquad O^-
\end{array}
$$

ethanolamine

phosphatidyl choline

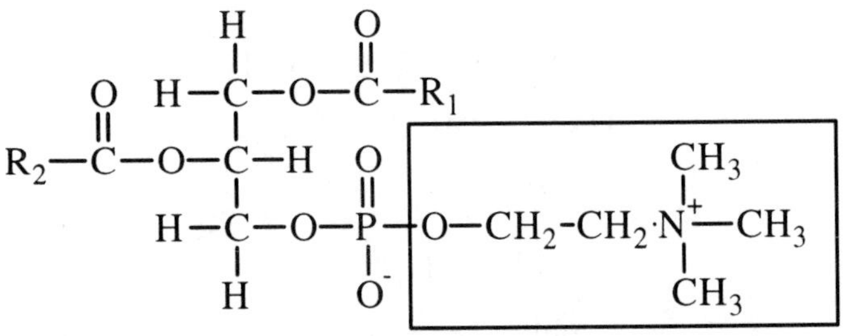

$$H-\overset{\overset{\displaystyle H}{|}}{C}-O-\overset{\overset{\displaystyle O}{\|}}{C}-R_1$$

$$R_2-\overset{\overset{\displaystyle O}{\|}}{C}-O-\overset{|}{C}-H$$

$$H-\overset{|}{C}-O-\overset{\overset{\displaystyle O}{\|}}{\underset{\underset{\displaystyle O^-}{|}}{P}}-O-CH_2-CH_2 \cdot \overset{+}{N}\begin{matrix} CH_3 \\ -CH_3 \\ CH_3 \end{matrix}$$

choline

phosphatidyl serine

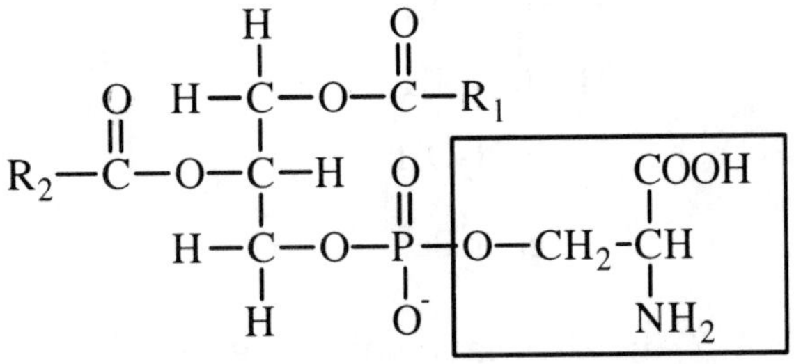

$$H-\overset{\overset{\displaystyle H}{|}}{C}-O-\overset{\overset{\displaystyle O}{\|}}{C}-R_1$$

$$R_2-\overset{\overset{\displaystyle O}{\|}}{C}-O-\overset{|}{C}-H$$

$$H-\overset{|}{C}-O-\overset{\overset{\displaystyle O}{\|}}{\underset{\underset{\displaystyle O^-}{|}}{P}}-O-CH_2-\overset{\overset{\displaystyle COOH}{|}}{\underset{\underset{\displaystyle NH_2}{|}}{CH}}$$

serine

phosphatidyl inositol

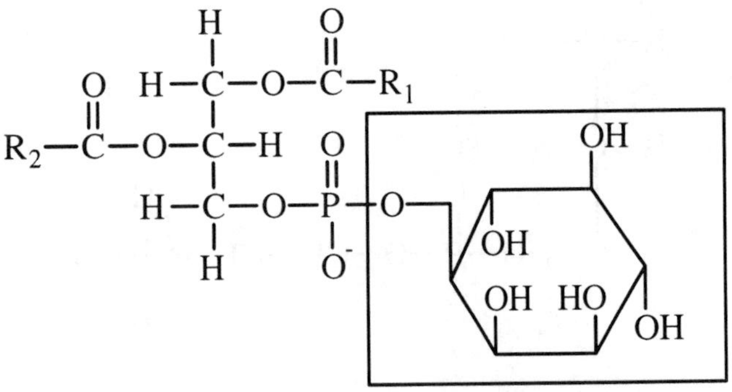

$$H-\overset{\overset{\displaystyle H}{|}}{C}-O-\overset{\overset{\displaystyle O}{\|}}{C}-R_1$$

$$R_2-\overset{\overset{\displaystyle O}{\|}}{C}-O-\overset{|}{C}-H$$

$$H-\overset{|}{C}-O-\overset{\overset{\displaystyle O}{\|}}{\underset{\underset{\displaystyle O^-}{|}}{P}}-O-$$

OH OH OH HO OH

inositol

Another lipid component of biological membranes is the **sphingolipids**. Sphingosine is the "backbone" of these lipids.

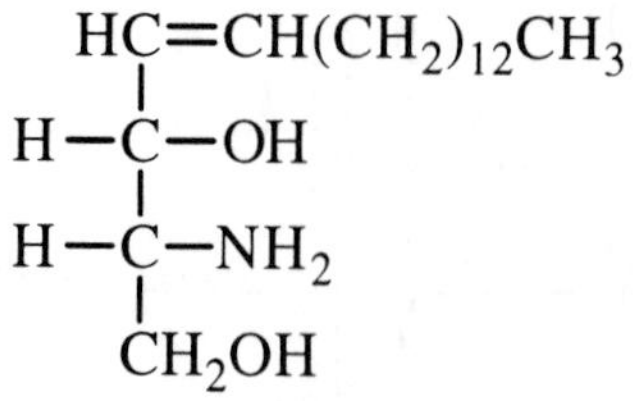

$$HC=CH(CH_2)_{12}CH_3$$
$$H-C-OH$$
$$H-C-NH_2$$
$$CH_2OH$$

sphingosine

When a fatty acid is attached to the amino group of sphingosine, the lipid is called a **ceramide**.

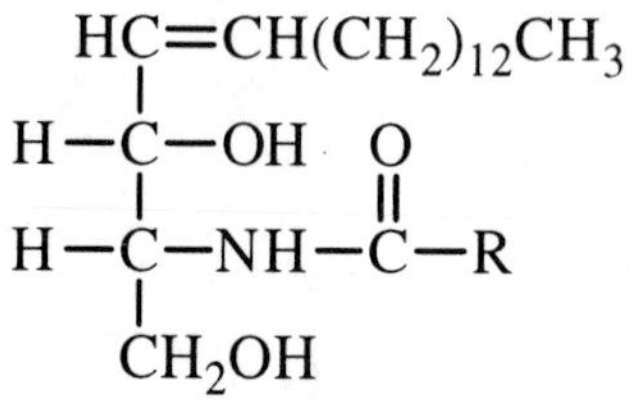

$$HC=CH(CH_2)_{12}CH_3$$
$$H-C-OH \quad O$$
$$H-C-NH-\overset{\shortparallel}{C}-R$$
$$CH_2OH$$

ceramide

The last hydroxyl group can also have substituents attached to it. Sphingomyelin, an important brain lipid, has a phosphocholine at the last position.

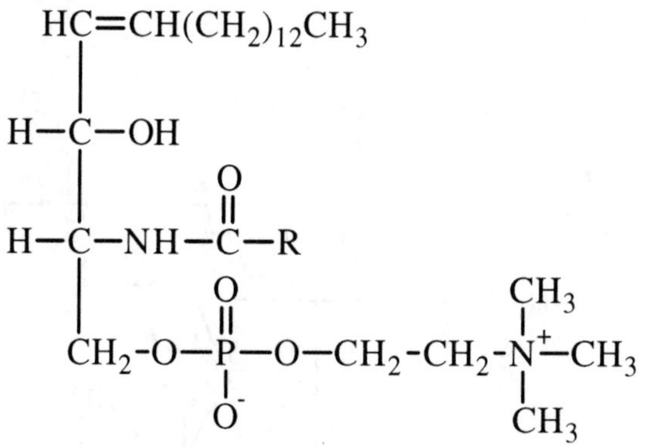

$$HC=CH(CH_2)_{12}CH_3$$
$$H-C-OH$$
$$H-C-NH-\overset{O}{\overset{\shortparallel}{C}}-R$$
$$CH_2-O-\overset{O}{\underset{O^-}{\overset{\shortparallel}{P}}}-O-CH_2-CH_2-\overset{CH_3}{\underset{CH_3}{\overset{|}{N^+}}}-CH_3$$

sphingomyelin

Sugars can also be found at the terminal hydroxyl of ceramide. These are generically called **glycolipids** (sugar + lipid).

Glucocerebroside and galactocerebroside are examples of glycolipids found in the brain.

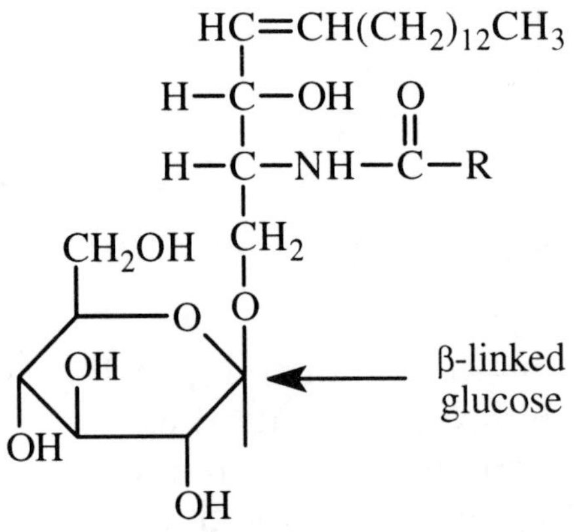

$$HC=CH(CH_2)_{12}CH_3$$

glucocerebroside

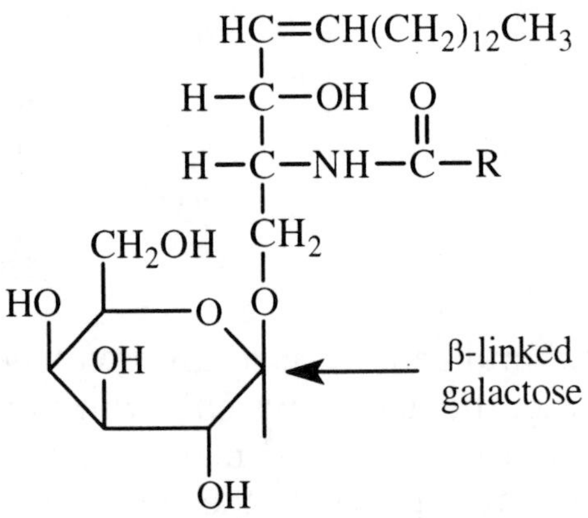

galactocerebroside

Isoprenoids are another important class of lipids. Its name is derived from the repeating isoprene unit from which they are made.

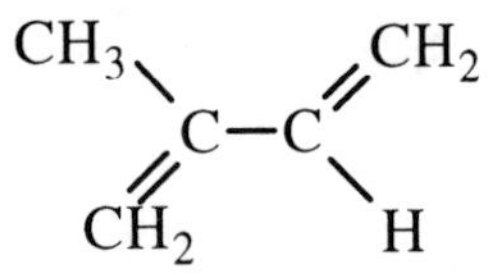

isoprene

Steroids are members of the isoprenoid family. Cholesterol is an example of a steroid found in the membranes of animals, but rarely in plants. That is why meats are high in cholesterol, whereas vegetables are not.

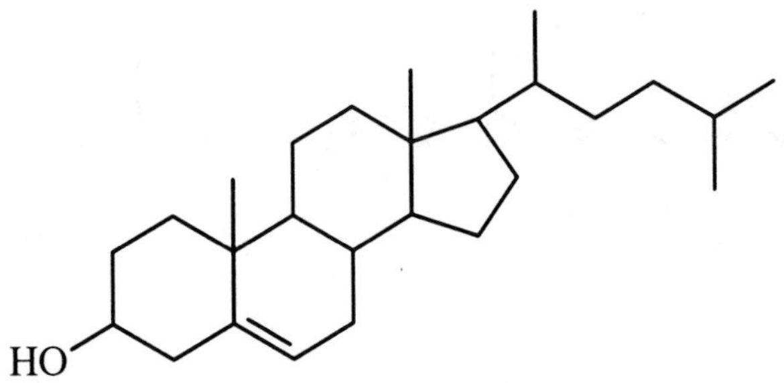

cholesterol

Many hormones and vitamins are also steroids. Testosterone, progesterone, and cortisol are examples of steroid hormones; vitamins A and D are examples of isoprenoid vitamins.

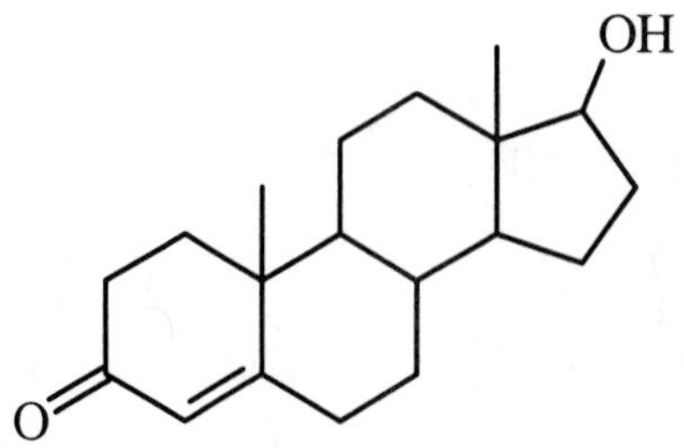

testosterone

Retinol and retinoic acid are involved in the regulation of cell growth. For example, the active ingredient in Retin-A™, which is both used as a treatment for acne and marketed as an anti-wrinkle cream, is retinoic acid. Retinoic acid prevents the synthesis of keratin, which is a tough skin protein.

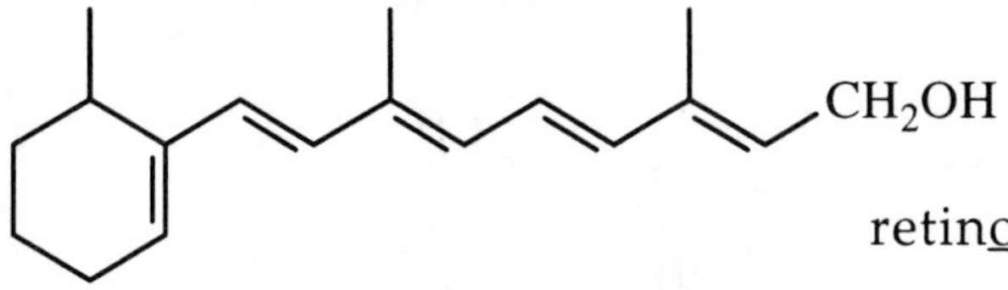

retinol

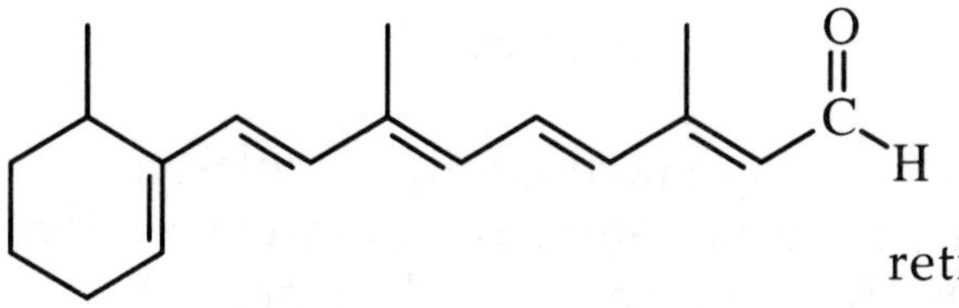

retinal

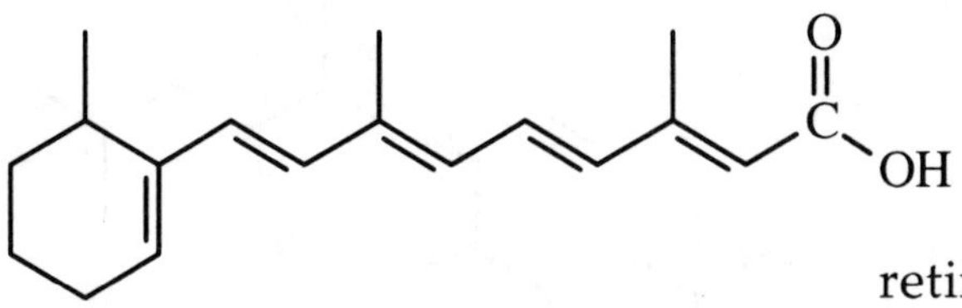

retinoic acid

The ability of our eyes to detect light involves the isoprenoid retinal. The double bonds in isoprenoids are generally in the *trans*-configuration. One of the double bonds in retinal (at position 11) becomes isomerized to the *cis*-configuration when light is absorbed. This structural change in retinal triggers a cascade of events which produce a neural signal in our brains.

11-*trans*-retinal $\xrightarrow{hv}$ 11-*cis*-retinal

Membrane structure and function

The structural characteristics of membranes are derived from the chemical properties of the major lipid components: phospholipids and sphingolipids.

At a particular concentration, known as the critical micellar concentration (cmc), lipid molecules come together to form spheres called micelles. In essence, the hydrophobic tails interact with one another to exclude water.

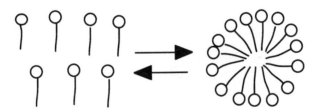

micelle formation

Phospholipids can also interact to form a bimolecular structure with two layers of lipids, called a **bilayer**.

This is the basic lipid structure of biological membranes:

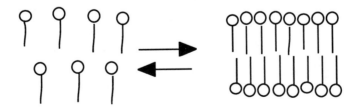

A bilayer is a good barrier to charged molecules like Na^+, K^+, Cl^-, and large polar molecules. However, small polar molecules like water can freely cross a lipid bilayer.

In addition to the lipid bilayer, biological membranes also require proteins. The function of membranes proteins are:

1. to allow molecules to cross the bilayer (transporters)
2. to allow signals to cross the bilayer (receptors)
3. to attach the membrane to other cellular components, such as the cytoskeleton

10 Lipid Metabolism

Objectives

1. Under what circumstances are triacylglycerols synthesized and degraded?

    When serum glucose levels are high, the hormone insulin promotes triacylglycerol synthesis (lipogenesis) by facilitating the transport of glucose to adipocytes. Adipocytes cannot synthesize triacylglycerols when glucose levels are low.

    When energy levels are low, the body's fat stores are mobilized in a process referred to as lipolysis. Several hormones stimulate the hydrolysis of triacylglycerols within the adipose tissue to form glycerol and fatty acids.

2. How are fatty acids degraded to generate energy?

    Most fatty acids are degraded to form acetyl-CoA within mitochondria in a process referred to as β-**oxidation**. Once acetyl-CoA is formed, it may then be oxidized in the citric acid cycle to generate energy.

3. What is the difference between fatty acid catabolic pathways referred to as β-oxidation and α-oxidation? Under what set of circumstances is each pathway used?

    The major difference between the two is that β-oxidation occurs at the β-carbon, whereas α-oxidation occurs at the α-carbon on a fatty acid.

    α-Oxidation occurs when there are branched and odd-chained fatty acids. The β-oxidation pathway is used in almost all other fatty acid degradation pathways.

4. What are ketone bodies? What purpose do they serve?

    In a process called ketogenesis, acetyl-CoA molecules are used to synthesize acetoacetate, β-hydroxybutyrate, and acetone, all of which belong to a group of molecules called the ketone bodies.

    Ketone bodies are used to generate energy in several tissues, most notably cardiac and skeletal muscle. During prolonged starvation, for example, the brain can also use ketone bodies as an energy source.

5. How are the eicosanoids synthesized? Why are specific drugs used to suppress eicosanoid synthesis?

The release of arachidonic acid, resulting from the binding of an appropriate chemical signal to its receptor on a target cell plasma membrane, is catalyzed by phospholipase A_2. Phospholipase A_2 cleaves acyl groups from C-2 of a phosphoglyceride, thus forming a fatty acid and lysophosphoglyceride. In some phosphoglycerides, the fatty acid is arachidonic acid, which can be converted into a variety of eicosanoids.

Specific drugs are used to suppress eicosanoid synthesis, resulting in the suppression of pain and inflammation in the tissue.

6. How and why do cells synthesize fatty acids?

Fatty acids are synthesized when the diet is low in fat and/or high in carbohydrate or protein. Most fatty acids are synthesized from dietary glucose. Glucose is converted to pyruvate in the cytoplasm. After entering the mitochondrion, pyruvate is converted to acetyl-CoA, which subsequently condenses with oxaloacetate to form citrate. When mitochondrial citrate levels are high, citrate enters the cytoplasm, where it is cleaved to form acetyl-CoA and oxaloacetate. In the presence of a large quantity of NADPH and ATP, a fatty acid can be synthesized from several acetyl-CoA.

7. How are membrane lipids synthesized and degraded? What is membrane remodeling?

The lipid bilayer of cell membranes is composed primarily of phospholipids and sphingolipids.

Phospholipid Synthesis

Once ethanolamine or choline has entered a cell, it is phosphorylated and converted to a CDP derivative. Subsequently, phosphatidyl ethanolamine or phosphatidyl choline is formed when diacylglycerol reacts with the CDP derivative. Triacylglycerol is produced if diacylglycerol reacts with acyl-CoA.

Phospholipid Degradation

Phospholipid degradation is catalyzed by several phospholipases.

Sphingolipid Synthesis

The synthesis of all sphingolipids begins with the production of ceramide. The synthesis of ceramide begins with the condensation of palmitoyl-CoA with serine to form 3-ketosphinganine. 3-Ketosphinganine is subsequently reduced by NADPH to form sphinganine. In a two-step process involving acyl-CoA and $FADH_2$, sphinganine is converted to ceramide. Ceramide can then be used to form sphingomyelin, glucosylceramide, or galactosylceramide.

Sphingolipid Degradation

Sphingolipids are degraded within lysosomes by specific hydrolytic enzymes.

Remodeling, a process that allows a cell to adjust the fluidity of its membranes, occurs when unsaturated fatty acids replace the original fatty acids incorporated during synthesis.

8. What similarities in isoprenoid synthetic pathway have been observed among living organisms?

Isoprenoids occur in all eukaryotes. Despite the astonishing diversity of isoprenoid molecules which are produced, there is a great deal of similarity in the synthesizing mechanism of different species. In fact, the initial phase of isoprenoid synthesis (the synthesis of isopentyl pyrophosphate) appears to be identical in all of the species in which this process has been investigated.

9. How is cholesterol synthesized and degraded? How is cholesterol metabolism regulated?

The synthesis of cholesterol can be divided into three phases:

a. formation of HMG-CoA (β-hydroxy-β-methylglutaryl-CoA) from acetyl-CoA,

b. conversion of HMG-CoA to squalene, and

c. conversion of squalene to cholesterol.

The most important mechanism for degrading and eliminating cholesterol is the synthesis of bile acids. The conversion of cholesterol to 7-α-hydrocholesterol, catalyzed by cholesterol-7-hydroxylase, is the rate-limiting reaction in bile acid synthesis. Subsequent reactions result in the rearrangement and reduction of double bond at C-5, the introduction of an additional hydroxyl group, and the reduction of the C-3-keto group to a 3-α-hydroxyl group. The products of this process, cholic acid and deoxycholic acid, are converted to bile salts by microsomal enzymes which catalyze conjunction reactions.

The amount and types of steroids synthesized in a specific tissue are carefully regulated. Cells in each tissue are programmed during embryonic and fetal development to respond to a variety of chemical signals by inducing the synthesis of a unique set of specific enzymes. The most important chemical signals that are now believed to influence steroid metabolism are various peptide hormones secreted from the pituitary and several prostaglandins.

General Principles

An adult human ingests 60-150g of fat per day. Ninety percent of this fat is triacylglycerols and 10% is phospholipids, cholesterol, and free fatty acids.

The first step in the processing of dietary fat is the hydrolysis of triacylglycerols to free fatty acids and monoacylglycerols. This is accomplished by gastric and pancreatic lipases.

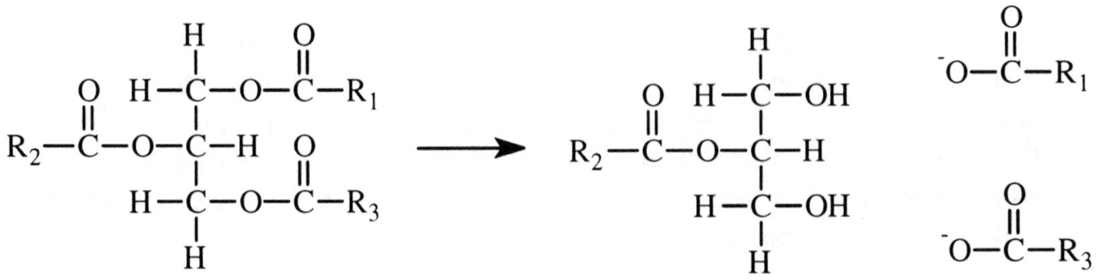

Free fatty acids have both polar and non-polar ends and can therefore function as biological "detergents" to solubilize triacylglycerols.

Further solubilization occurs in the intestine with bile salts. Bile salts, like sodium cholate, are biological detergents made by the liver and secreted into the small intestine.

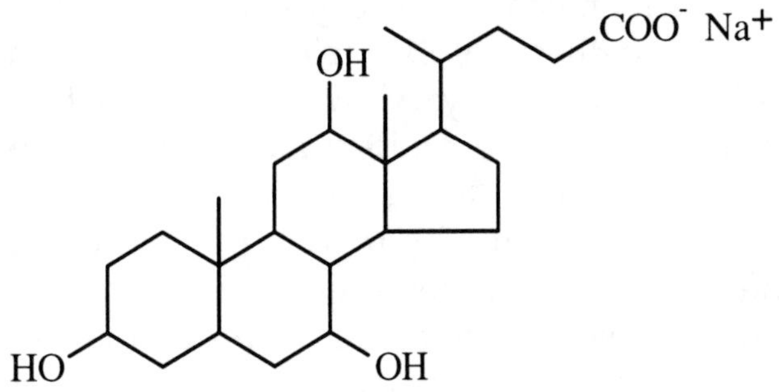

Sodium Cholate

Bile salts form mixed micelles with other fats. Micelles are then taken up into intestinal cells and transported in the blood.

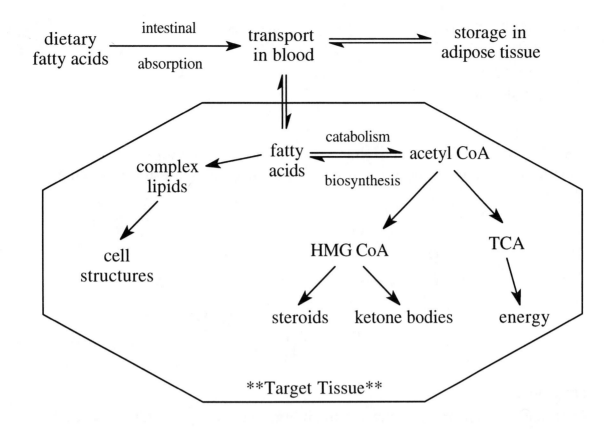

Target tissues:

The use of fatty acids for energy depends on the tissue. Nervous tissue, for instance, uses fatty acids minimally, but fatty acids are a major source of energy for muscle. During fasting, many tissues are able to use fatty acids or ketone bodies for energy.

Fatty acid catabolism:
- Activation
- Transport
- β-oxidation

Activation

Acyl-CoA synthetase is an enzyme associated with the endoplasmic reticulum, or outer mitochondrial membrane, and catalyzes the following reaction:

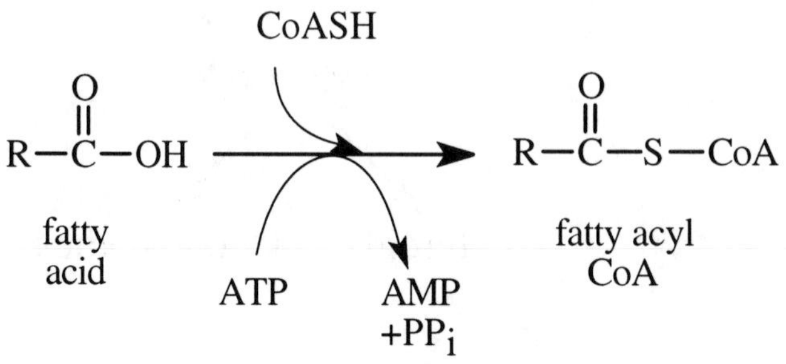

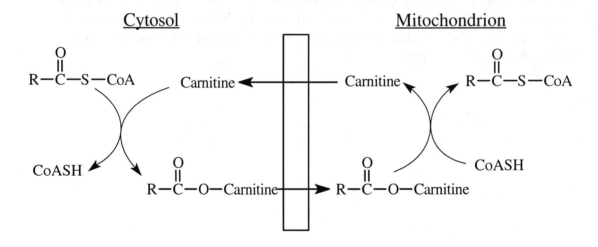

Transport

Fatty acid catabolism occurs inside the mitochondrion. Since fatty acyl CoA cannot cross the membrane, the fatty acyl group is transferred to carnitine, which can be transported into the mitochondrion. The cell needs to maintain separate cytosolic and mitochondrial pools of CoASH.

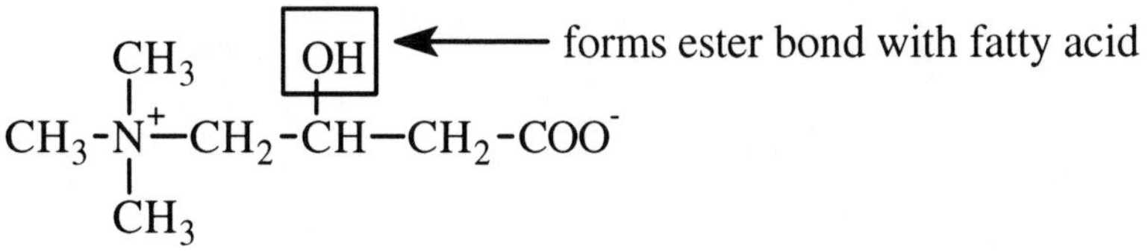

<u>β-oxidation</u>

Occurs in four steps:
- oxidation (remove electrons)
- hydration (add water)
- oxidation (remove more electrons)
- thiolysis (remove acetyl-CoA)

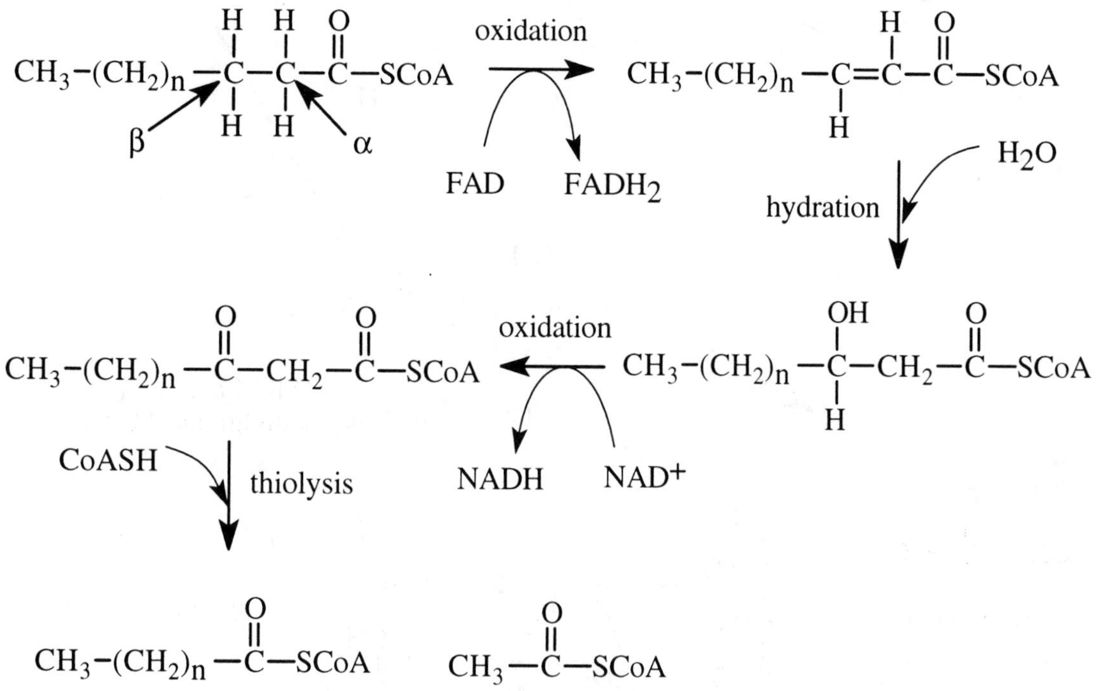

$$CH_3-(CH_2)_n-\underset{\underset{\beta}{\overset{|}{H}}}{\overset{\overset{H}{|}}{C}}-\underset{\underset{\alpha}{\overset{|}{H}}}{\overset{\overset{H}{|}}{C}}-\overset{\overset{O}{\|}}{C}-SCoA \xrightarrow{\text{oxidation}} CH_3-(CH_2)_n-\overset{\overset{H}{|}}{C}=\underset{\underset{H}{|}}{C}-\overset{\overset{O}{\|}}{C}-SCoA$$

FAD → FADH₂ (oxidation step)

hydration + H_2O

$$CH_3-(CH_2)_n-\overset{\overset{O}{\|}}{C}-CH_2-\overset{\overset{O}{\|}}{C}-SCoA \xleftarrow{\text{oxidation}} CH_3-(CH_2)_n-\underset{\underset{H}{|}}{\overset{\overset{OH}{|}}{C}}-CH_2-\overset{\overset{O}{\|}}{C}-SCoA$$

NADH ← NAD⁺ (oxidation step)

CoASH → thiolysis

$$CH_3-(CH_2)_n-\overset{\overset{O}{\|}}{C}-SCoA \qquad CH_3-\overset{\overset{O}{\|}}{C}-SCoA$$

Each round of β-oxidation produces one FADH$_2$, one NADH, and one acetyl-CoA. FADH$_2$ and NADH enter the electron transport chain, which generates a proton gradient for ATP synthesis. The acetyl-CoA molecule can be further oxidized in the citric acid cycle to produce three NADH, one FADH$_2$, and one GTP.

<u>Ketone Bodies</u>

Ketone bodies function as metabolic fuel for many peripheral tissues, particularly the heart and skeletal muscle. Ketone bodies can be thought of as a water-soluble, fatty acid equivalent.

In the liver, acetyl-CoA can be converted into ketone bodies (acetoacetate and β-hydroxybutyrate) by a process known as **ketogenesis**.

Biosynthesis of ketone bodies (ketogenesis):

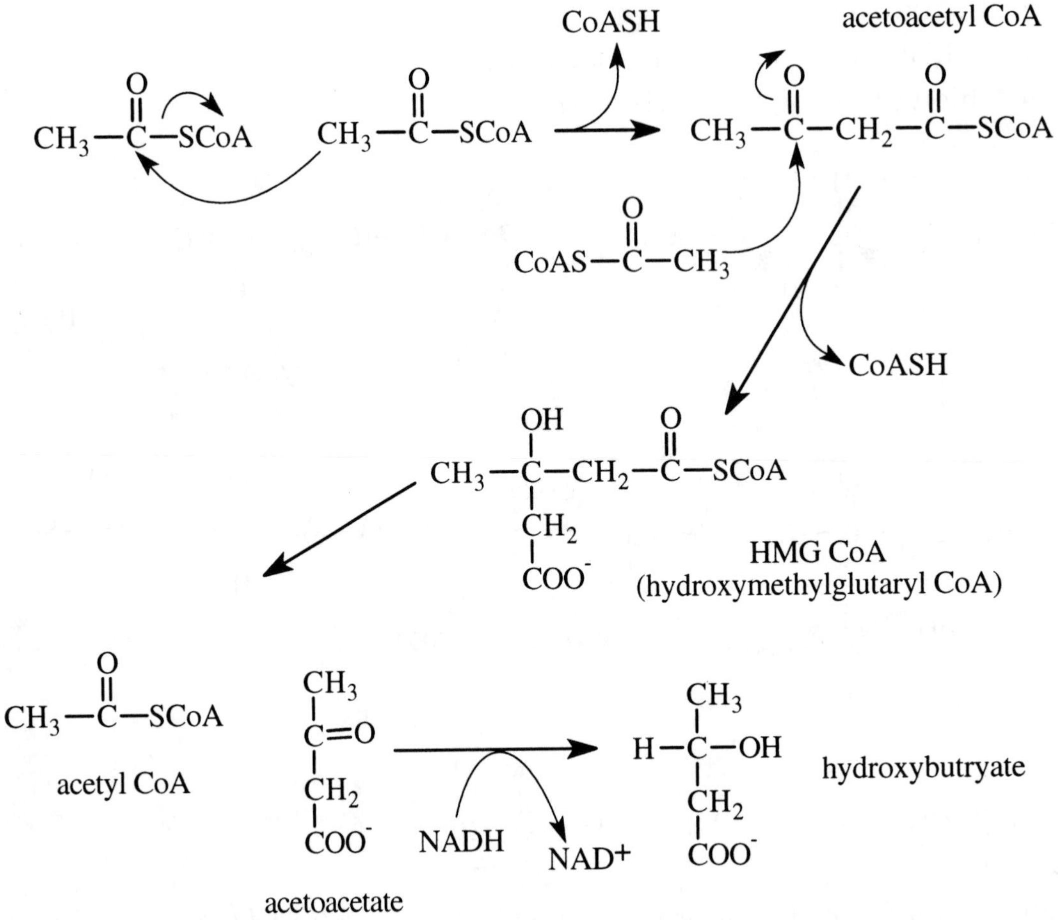

Overall reaction:

$$2 \text{ acetyl-CoA} \longrightarrow \text{acetoacetate} + 2 \text{ CoASH}$$

Note: It takes a total of three acetyl-CoA molecules to form one acetoacetate molecule. Actually, one acetyl-CoA is recycled in the synthesis of ketone bodies and can therefore be considered catalytic.

The liver releases acetoacetate and β-hydroxybutyrate into the blood for other tissues to use.

In the target tissues, acetoacetate is converted back into two molecules of acetyl-CoA.

80

In tissues:

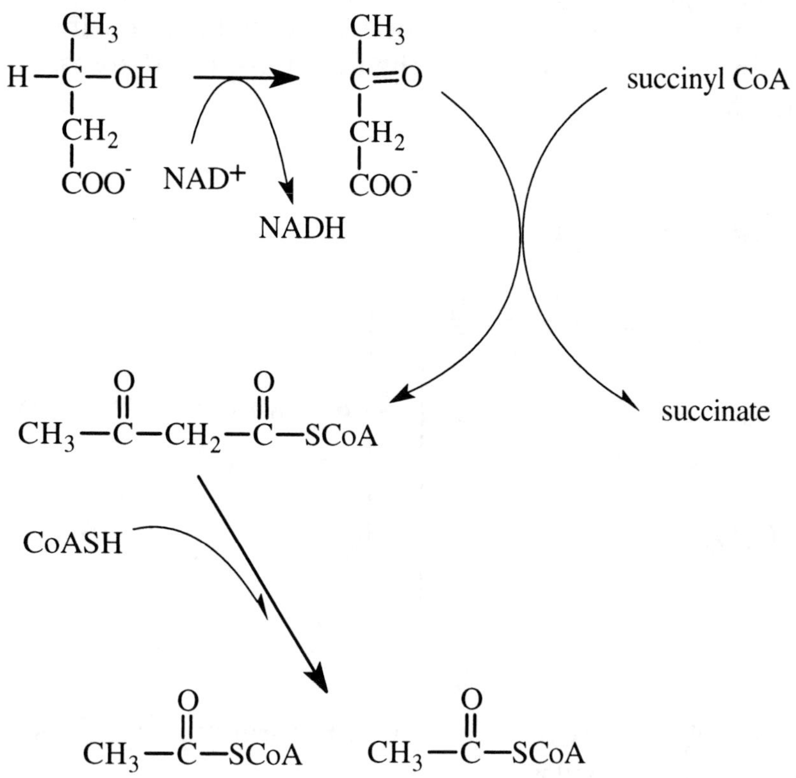

Ketosis is a pathological condition which occurs when acetoacetate is produced faster than it can be metabolized. Under these conditions, the non-enzymatic degradation of acetoacetate occurs:

$$CH_3-\overset{\overset{O}{\|}}{C}-CH_2\text{-}COO^- \longrightarrow CH_3\text{-}\overset{\overset{O}{\|}}{C}-CH_3 \quad CO_2$$

Fatty acid biosynthesis is not a simple reversal of β-oxidation because of thermodynamic considerations.

Acetyl-CoA is produced in the mitochondrion and must be transported to the cytosol, but once again, CoASH is unable to cross the membrane and so an alternative pathway is necessary. This occurs by first condensing acetyl-CoA with OAA to form citrate. Citrate can then be transported across the mitochondrial membrane.

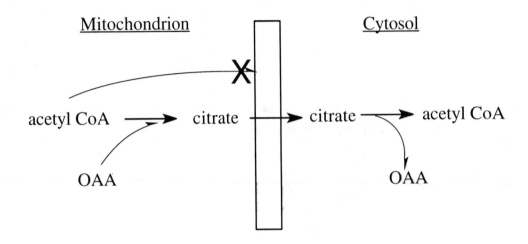

Once in the cytosol, acetyl-CoA is activated to form malonyl-CoA by the enzyme acetyl-CoA carboxylase complex.

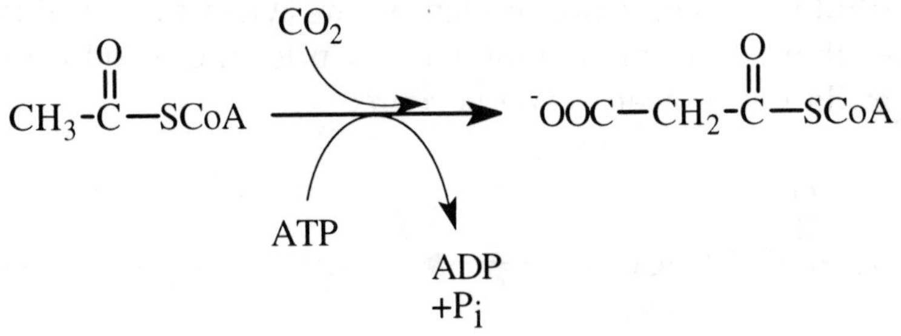

$$CH_3-\overset{\overset{\displaystyle O}{\|}}{C}-SCoA \xrightarrow{\hspace{2cm}} {}^-OOC-CH_2-\overset{\overset{\displaystyle O}{\|}}{C}-SCoA$$

CO$_2$

ATP

ADP
+P$_i$

The malonyl-CoA group is transferred from CoASH to an acyl carrier protein (ACP). The malonyl-ACP serves as the carbon donor for fatty acid biosynthesis.

Summary of fatty acid biosynthesis:

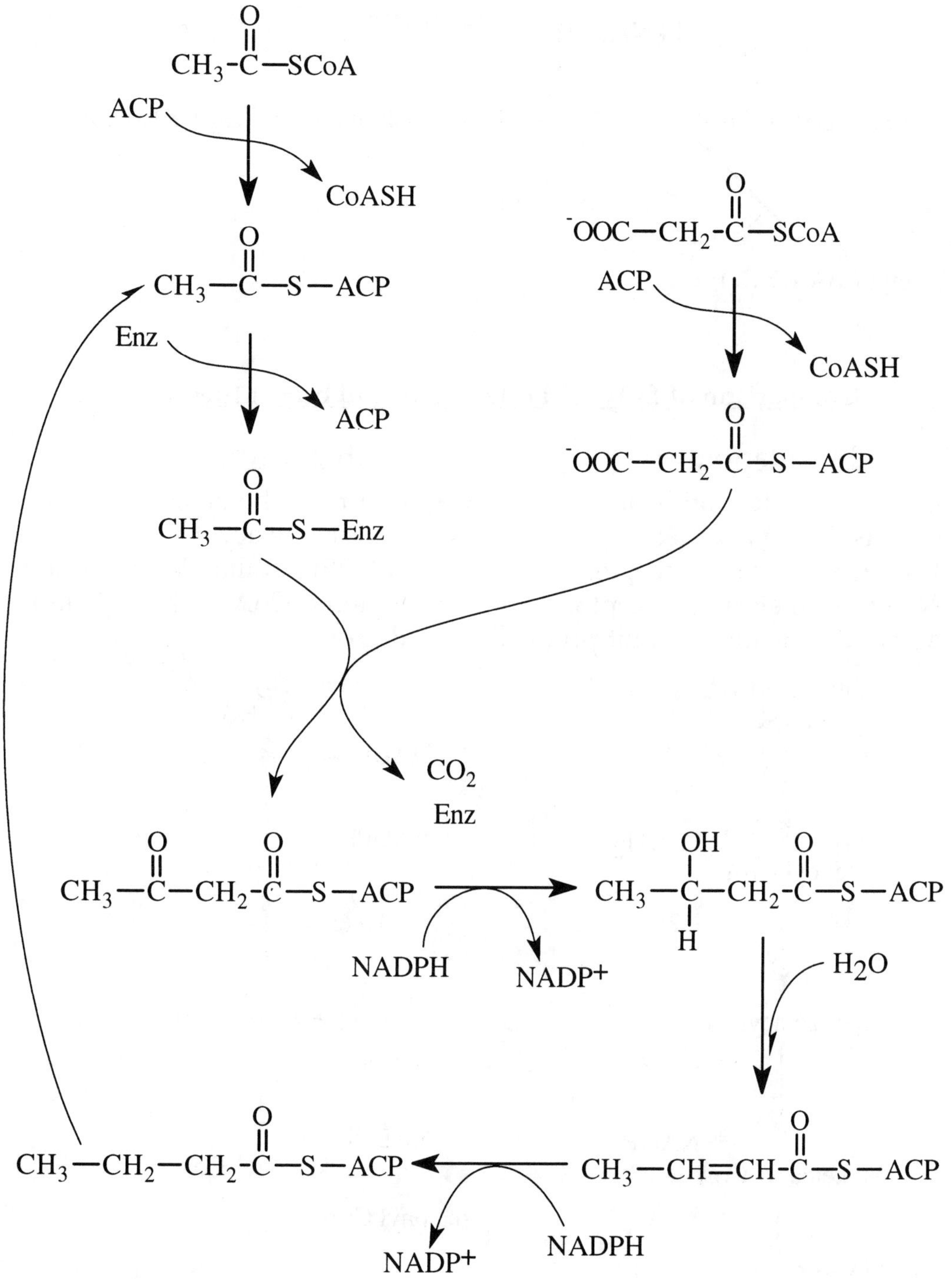

$$CH_3-\overset{\overset{\displaystyle O}{\|}}{C}-SCoA$$

ACP → CoASH

$$CH_3-\overset{\overset{\displaystyle O}{\|}}{C}-S-ACP$$

Enz → ACP

$$CH_3-\overset{\overset{\displaystyle O}{\|}}{C}-S-Enz$$

$$^-OOC-CH_2-\overset{\overset{\displaystyle O}{\|}}{C}-SCoA$$

ACP → CoASH

$$^-OOC-CH_2-\overset{\overset{\displaystyle O}{\|}}{C}-S-ACP$$

→ CO_2

Enz

$$CH_3-\overset{\overset{\displaystyle O}{\|}}{C}-CH_2-\overset{\overset{\displaystyle O}{\|}}{C}-S-ACP \longrightarrow CH_3-\overset{\overset{\displaystyle OH}{|}}{\underset{\underset{\displaystyle H}{|}}{C}}-CH_2-\overset{\overset{\displaystyle O}{\|}}{C}-S-ACP$$

NADPH → NADP⁺

H_2O

$$CH_3-CH_2-CH_2-\overset{\overset{\displaystyle O}{\|}}{C}-S-ACP \longleftarrow CH_3-CH=CH-\overset{\overset{\displaystyle O}{\|}}{C}-S-ACP$$

NADP⁺ ← NADPH

The overall reaction to synthesize a C-16 fatty acid is:

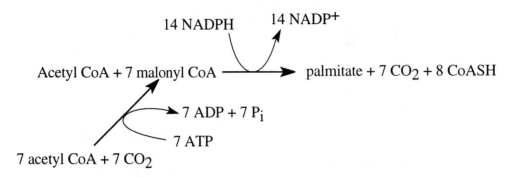

Comparison of fatty acid catabolism and biosynthesis

β-Oxidation

- occurs in mitochondrion
- CoA is the acyl carrier
- FAD is an electron acceptor
- NAD⁺ is an electron acceptor
- acetyl-CoA is the C-2 unit product

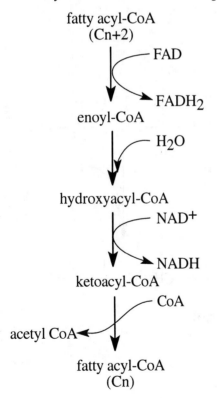

Biosynthesis

- occurs in the cytosol
- ACP is the acyl carrier
- NADPH is the electron donor
- malonyl-CoA is the C-2 unit donor

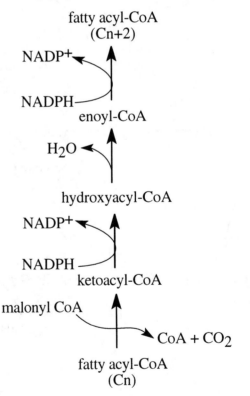

11   Aerobic Metabolism

Objectives

1. How is energy obtained from the breakdown of acetyl-CoA by the reactions of the citric acid cycle?

    The citric acid cycle is a series of biochemical reactions which are responsible for the eventual oxidation of acetyl-CoA into CO_2 and H_2O. The electrons which are removed from the citric acid cycle intermediates are transferred to NAD^+ and FAD to form the reduced coenzymes NADH and $FADH_2$. Energy is then captured as ATP during the reoxidation of NADH and $FADH_2$ via oxidative phosphorylation.

2. How is the energy released during the electron transport pathway captured and used to drive biosynthetic processes?

    The electron transport chain is a series of electron carriers which transfer the electrons from the reduced coenzymes to oxygen. The energy released during electron transfer is coupled to several endergonic processes, the most prominent of which is ATP synthesis. The hydrolysis of ATP, therefore, is used to drive biosynthetic processes.

3. How are the pathways of energy metabolism interrelated?

    The molecules of NADH and $FADH_2$ which are produced in glycolysis and the citric acid cycle are used to generate usable energy in the electron transport pathway. The pathway consists of a series of redox carriers which receive electrons from NADH and $FADH_2$. At the end of the pathway the electrons, along with protons, are donated to oxygen to form H_2O.

4. How are the pathways of energy metabolism regulated so that each cell's energy requirements are consistently met?

    Regulation is achieved primarily by modulation of key enzymes and the availability of certain substrates. Because of its prominent role in energy production, the cycle is also dependent on a continuous supply of NAD^+, FAD, and ADP. The citric acid cycle enzymes citrate synthase, isocitrate dehydrogenase, and α-ketoglutarate dehydrogenase, are highly regulated because they catalyze reactions which represent important metabolic branch points.

    Because the concentrations of acetyl-CoA and oxaloacetate are low in mitochondria in relation to the amount of the enzyme, any increase in substrate availability stimulates citrate synthesis. Therefore the rate of citrate synthesis is influenced by changes in concentrations of acetyl-CoA and oxaloacetate. High concentrations of succinyl-CoA and citrate inhibit citrate synthase by acting as allosteric inhibitors.

Isocitrate dehydrogenase catalyzes the second highly regulated reaction in the cycle. Its activity is stimulated by relatively high concentrations of ADP and NAD^+ and inhibited by ATP and NADH. An important reason for the high degree of regulation of isocitrate dehydrogenase is its role in lipid metabolism.

The activity of α-ketoglutarate dehydrogenase is carefully regulated because of the important role of α-ketoglutarate in several metabolic processes. When a cell's energy stores are low, the activation of α-ketoglutarate dehydrogenase results in the retention of α-ketoglutarate within the cycle at the expense of biosynthetic processes. As the cell's supply of NADH rises, the enzyme is inhibited and α-ketoglutarate molecules become available for biosynthetic reactions.

5. Why do calorie-restrictive diets promote obesity?

When individuals diet, their brains seem to interpret lower food intake as the onset of famine. In an attempt to prevent starvation, the body adapts by reducing its basal metabolic rate. Initial weight loss which results from calorie restriction is actually a loss of muscle. Because of its high consumption of ATP during muscle contraction and its relatively large mass, skeletal muscle is a metabolically demanding tissue. If muscle mass is reduced, the body can conserve vital energy reserves which can be used to outlive the famine. Therefore, calorie-restrictive diets can result in obesity by reducing skeletal muscle tissue and weight. Weight is then regained in the form of fat in adipose tissue.

6. What are the toxic products of oxygen metabolism and how do they damage cells?

A number of phenomena including cancer, myocardial infarct, inflammation, and aging, have been associated with the formation of oxygen derivatives referred to as reactive oxygen species (ROS). Examples of ROS include the superoxide radical, hydrogen peroxide, hydroxyl radical, and singlet oxygen. Damage results primarily from enzyme inactivation, polysaccharide depolymerization, DNA breakage, and membrane destruction. Examples of circumstances which may cause serious oxidative damage include overconsumption of certain drugs or exposure to certain environmental contaminants.

7.   How are toxic oxygen metabolites formed and destroyed?

During mitochondrial electron transport, H_2O is formed as a consequence of the sequential transfer of four electrons to O_2. During this process, several ROS are created. ROS may occasionally leak out of active sites before they can be completely reduced. Under normal circumstances, cellular antioxidant defense mechanisms minimize any subsequent damage. ROS are also formed during non-enzymatic processes. For example, exposure to UV light and ionizing radiation can result in ROS formation.

The major enzymatic defenses against oxidative stress are provided by superoxide dismutase, catalase, and glutathione peroxidase. Superoxide dismutases are a class of enzymes which catalyze the formation of H_2O_2 and O_2 from the superoxide radical. Catalase is a heme-containing enzyme that uses H_2O_2 to oxidize other substrates. When H_2O_2 is present in excessive amounts, catalase converts it into water. Glutathione peroxidase is a key component in an enzymatic system that is most responsible for controlling cellular peroxide levels. In addition to reducing hydrogen peroxide to form water, glutathione peroxidases transform organic peroxides into alcohols.

Living organisms utilize a variety of antioxidant molecules to protect themselves from radicals. Some prominent antioxidants include glutathione (GSH), α-tocopherol (vitamin E), ascorbic acid (vitamin C), and β-carotene. α-Tocopherol, a potent radical scavenger, belongs to a class of compounds referred to as phenolic antioxidants. Phenols are effective antioxidants because the radical products of these molecules are resonance-stabilized and thus relatively stable. Because vitamin E is lipid-soluble, it plays an important role in protecting membranes from lipid peroxyl radicals. β-Carotene is a member of a class of plant pigment molecules referred to as the carotenoids. In plant tissue the carotenoids absorb some of the light energy used to drive photosynthesis and protect against the ROS that are created at high light intensities. In animals, β-carotene is a precursor of retinol and an important antioxidant in membranes. Ascorbic acid has been shown to be an efficient antioxidant. The water-soluble form of ascorbic acid, ascorbate, scavenges a variety of ROS within the aqueous compartments of cells and in extracellular fluids. Ascorbate protects membranes via two mechanisms. First, ascorbate reacts with peroxyl radicals formed in the cytoplasm before they can reach the membrane, thereby preventing lipid peroxidation. Second, ascorbate enhances the antioxidant activity of vitamin E by generating reduced α-tocopherol from the α-tocopheroloxyl radical.

Study Tips

In Chapter 8 of this study guide, we discussed that the fate of pyruvate depends on whether oxygen is present. We will now follow the fate of pyruvate.

Once again, think **function**.

The function of the citric acid cycle is to:
1. produce intermediates
2. produce energy

Energy (in the form of ATP) is not explicitly produced by the citric acid cycle. The overall reaction is that for one pyruvate molecule (three carbons) entering the cycle, three CO_2 are produced as well as one molecule of GTP, four molecules of NADH and one molecule of $FADH_2$.

The intermediates in the citric acid cycle are a rich source of starting material for numerous biosynthetic pathways. The carbon precursors for lipids, sugars, amino acids, and nucleic acids are all derived from citric acid cycle intermediates.

However, if pyruvate (three carbons) enters and three CO_2 leave, how does the citric acid cycle produce intermediates?

Possible solutions:

There must be either:
1. another way for carbon to enter the cycle (other than via acetyl-CoA),

or

2. a way to bypass the CO_2 generating steps of the cycle.

Both possibilities exist, but it depends on the organism as to which one is used.

Mammals use the first option. Rather than sending the carbons of pyruvate through acetyl-CoA and on into the cycle (which does not result in the net increase in cycle intermediates), pyruvate enters the cycle through an alternative route.

Pyruvate is converted directly to oxaloacetate by the enzyme pyruvate carboxylase.

$$
\begin{array}{c}
COO^- \\
| \\
C=O \\
| \\
CH_3
\end{array}
\quad
\xrightarrow[\text{ATP} \quad \text{ADP} + P_i]{\text{CO}_2}
\quad
\begin{array}{c}
COO^- \\
| \\
C=O \\
| \\
CH_2 \\
| \\
COO^-
\end{array}
$$

This reaction results in the <u>net</u> increase in the concentration of citric acid cycle intermediates.

Plants use the second option to make citric acid cycle intermediates. They bypass the CO_2 generating steps by siphoning away isocitrate. This option, the glyoxylate bypass, or shunt, is comprised of two enzymatic steps. In essence, two acetyl-CoA molecules are converted into one malate molecule.

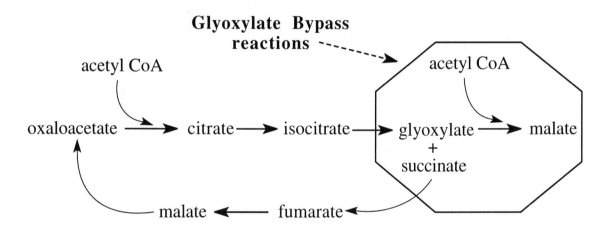

Cytosolic NADH is generated by glycolysis in the conversion of glucose to pyruvate (refer to Chapter 8). Remember, NADH needs to be recycled back to NAD^+ for glycolysis to proceed. In the absence of oxygen, NADH $\longrightarrow$ NAD^+ by lactate dehydrogenase. In the presence of oxygen, pyruvate can enter the citric acid cycle.

What happens to the cytosolic NADH in the presence of oxygen?

The electrons from NADH are shuttled into the mitochondrion for ATP synthesis. Since NADH cannot cross the mitochondrial membrane, the electrons associated with NADH must be transferred to a compound that can cross the mitochondrial membrane.

Two shuttle mechanisms exist for the regeneration of NADH and the transfer of electrons into the mitochondrion. They are: 1) glycerol phosphate and 2) malate-aspartate.

1. Glycerol Phosphate. Electrons are transferred from NADH to dihydroxyacetone phosphate (DHAP) to form glycerol phosphate. The electrons are then transferred to FAD in the mitochondrion.

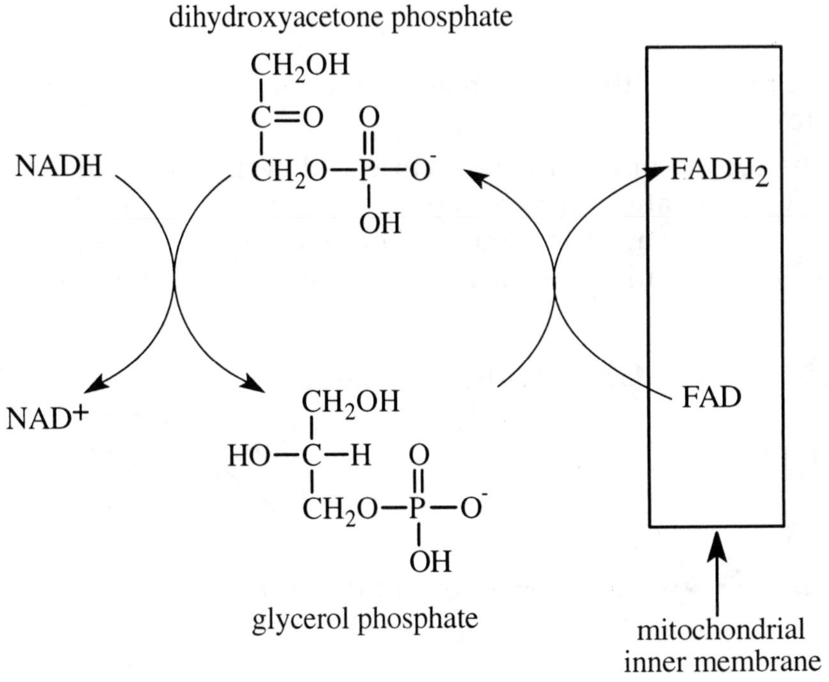

dihydroxyacetone phosphate

$$CH_2OH$$
$$C=O$$
$$CH_2O-P-O^-$$
$$OH$$

NADH

NAD⁺

$$CH_2OH$$
$$HO-C-H$$
$$CH_2O-P-O^-$$
$$OH$$

glycerol phosphate

FADH$_2$

FAD

mitochondrial inner membrane

2. Malate-Aspartate. Electrons are transferred from NADH to oxaloacetate (OAA) to form malate. Malate then crosses the mitochondrial membrane and the electrons are released inside to form NADH. Inside the mitochondrion, OAA is converted into the amino acid aspartate, which is then transported out of the mitochondrion. Aspartate is then converted back to OAA, thereby completing the cycle.

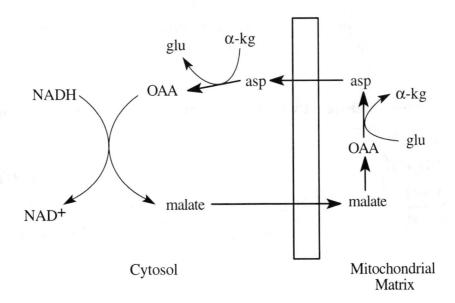

Cytosol   Mitochondrial
                              Matrix

Pentose phosphate pathway

Think function, function, function. Although the "oxidative" and "non-oxidative" reactions of the pentose phosphate pathway are generally grouped together, it is sometimes easier if you consider each phase separately.

Two-fold function:

1. Produce reducing power (a source of electrons) for biosynthesis.
2. Produce sugar intermediates for biosynthesis.

The production of reducing power is accomplished with the first three reactions of the pentose phosphate pathway, in which NADPH is generated as a source of electrons.

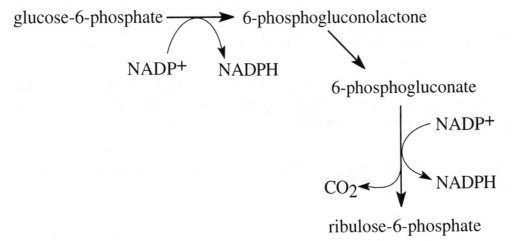

These reactions are essentially irreversible.

The second function – production of sugar intermediates – is accomplished by the transaldolase (3-carbon transfer) and transketolase (2-carbon transfer) reactions.

Transketolase: Transfers two carbons from a keto sugar to an aldo sugar

Keto sugar   Aldo sugar

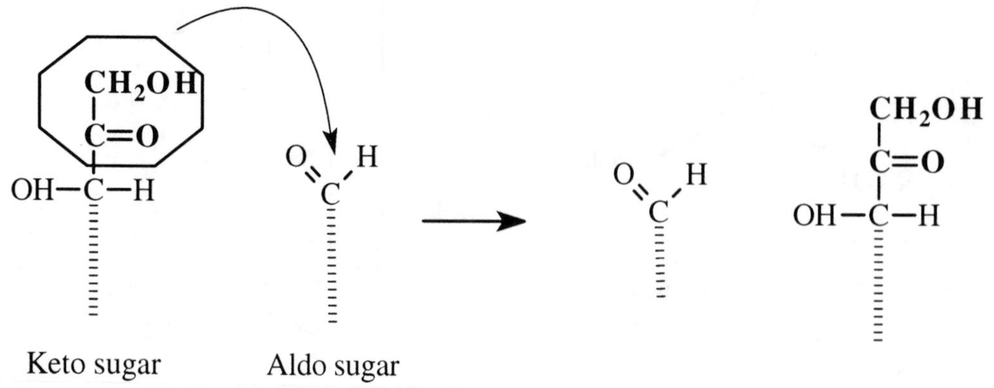

Transaldolase: Transfers three carbons from a keto sugar to an aldo sugar

Keto sugar   Aldo sugar

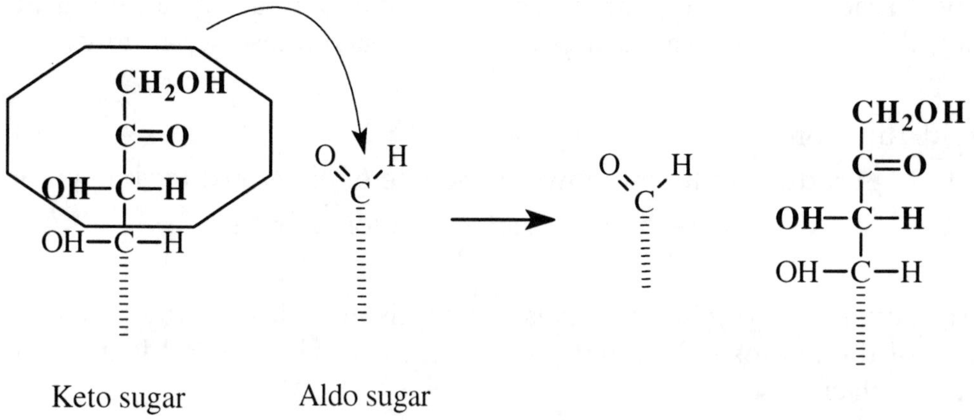

These reactions are completely reversible.

After glucose-6-phosphate is oxidized to generate two NADPH molecules (and one CO_2 molecule), the remaining carbons become ribulose-5-phosphate. What happens to these carbons?

The answer to this question depends on the metabolic needs of the cells. (Hint: think function!)

If the cell needs sugars for nucleotide biosynthesis, the ribulose-5-phosphate molecule will be isomerized to ribose-5-phosphate and then used to make nucleotides.

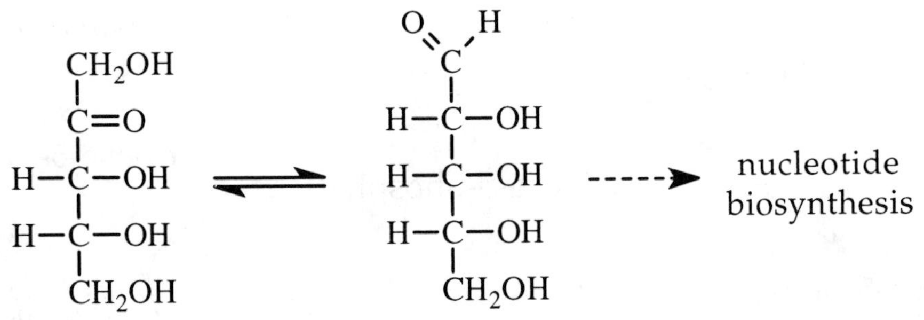

$$
\begin{array}{ccc}
\begin{array}{c}
CH_2OH \\
| \\
C=O \\
| \\
H-C-OH \\
| \\
H-C-OH \\
| \\
CH_2OH
\end{array}
& \rightleftharpoons &
\begin{array}{c}
O_{\diagdown}\;{}^{H}\!\!\diagup \\
C \\
| \\
H-C-OH \\
| \\
H-C-OH \\
| \\
H-C-OH \\
| \\
CH_2OH
\end{array}
& - - - \rightarrow &
\begin{array}{c}
\text{nucleotide} \\
\text{biosynthesis}
\end{array}
\end{array}
$$

ribulose-5-phosphate   ribose-5-phosphate

However, if the cell only needs the pentose phosphate pathway for reducing equivalents (NADPH), then the carbons of ribulose-5-phosphate will be recycled and converted into intermediates of glycolysis, namely fructose-6-phosphate and glyceraldehyde-3-phosphate. This is accomplished by a series of transaldolase (3-carbon transfer) and transketolase (2-carbon transfer) reactions.

In order to completely recycle the ribulose-5-phosphate carbons into glycolysis, at least three molecules of ribulose-5-phosphate are required initially (see Figure 11-1).

What if the cell needs ribose-5-phosphate but does not need reducing power (NADPH)?

The cell would use the transaldolase and transketolase reactions, starting with fructose-6-phosphate and glyceraldehyde-3-phosphate (see Fig. 11-2). The transaldolase, transketolase, epimerase, and isomerase reactions are reversible ($\Delta G=0$).

Figure 11-1

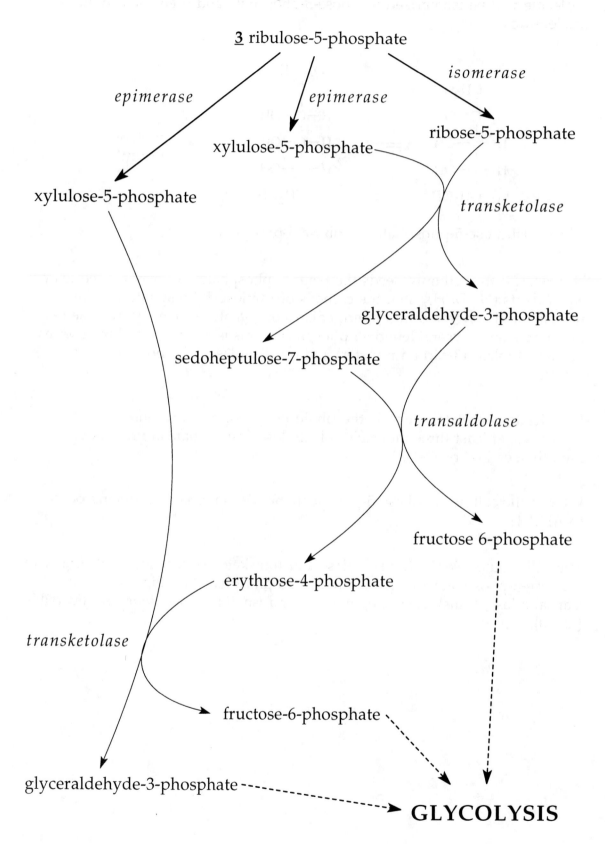

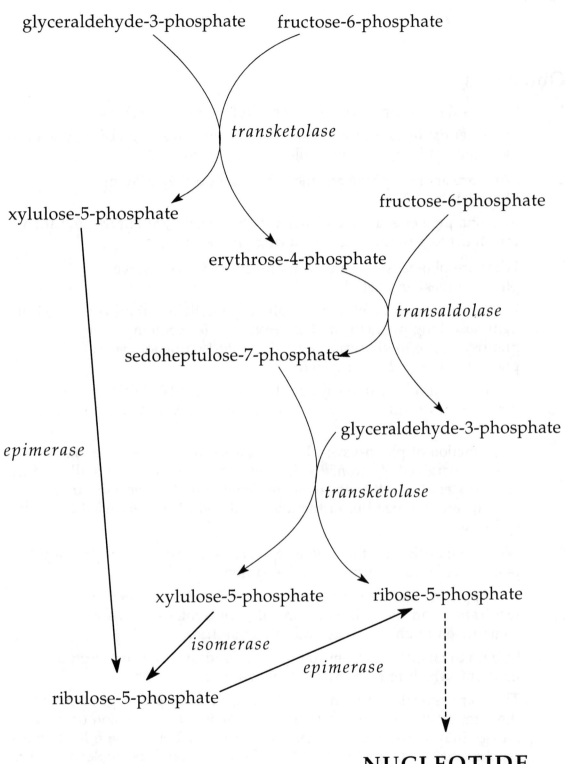

Figure 11-2

glyceraldehyde-3-phosphate   fructose-6-phosphate

transketolase

xylulose-5-phosphate

fructose-6-phosphate

erythrose-4-phosphate

transaldolase

sedoheptulose-7-phosphate

glyceraldehyde-3-phosphate

epimerase

transketolase

xylulose-5-phosphate   ribose-5-phosphate

isomerase

epimerase

ribulose-5-phosphate

**NUCLEOTIDE
BIOSYNTHESIS**

12 Photosynthesis

Objectives

1. What is the principal role of chlorophyll in photosynthesis?

   The essential feature of photosynthesis is the absorption of light energy by specialized pigment molecules called chlorophyll.

2. What are antenna pigments and what role do they play in photosynthesis?

   Antenna pigments are molecules which absorb light energy and then transfer it to a reaction center during photosynthesis.

3. What are photosystems and what function do they serve in photosynthesizing cells?

   Photosystems are a collection of photosynthetic reactions composed of light-absorbing pigments and electron transfer proteins. The photosynthetic mechanism is divided into two major parts: photosystem I and photosystem II.

   The essential role of photosystem I is to produce NADPH by the donation of energized electrons to a series of electron carriers within the thylakoid membrane.

   The function of photosystem II is to oxidize water molecules and donate energized electrons to electron carriers which eventually reduce photosystem I. As the electrons are donated to the electron carriers, protons are pumped out of the thylakoid and ultimately used for ATP synthesis.

4. What similarities are there between aerobic respiration and the light reactions of photosynthesis? What differences exist?

   It is apparent that the differences and similarities between aerobic respiration and the light reactions of photosynthesis involve a comparison of chloroplasts and mitochondria.

   In both organelles, electron transport is used to produce a proton gradient, which in turn drives ATP synthesis.

   There are essentially two differences between aerobic respiration and the light reactions of photosynthesis. One is the conversion of light energy into redox energy by the chloroplasts, whereas the mitochondria produce redox energy by extracting electrons from food molecules. The other critical difference involves the permeability characteristics of mitochondrial inner membrane and thylakoid membrane. In contrast to the inner membrane, the thylakoid membrane is permeable to Mg^+ and Cl^- as well as other ions.

5. What is the Z scheme?

The Z scheme is a mechanism by which electrons flow between photosystem II and photosystem I during photosynthesis.

6. What is the water-oxidizing clock?

The water-oxidizing clock is the mechanism by which H_2O is converted into O_2.

7. What is the metabolic relationship between the light reactions and the light-independent reactions of photosynthesis?

The light-independent reactions, also known as dark reactions, are related to the light reactions through ATP and NADPH. The fixation of CO_2 by the Calvin cycle cannot occur without sufficient ATP and NADPH. Under normal conditions, approximately 30% of the CO_2 fixed by the leaves is incorporated into starch, which is stored as water-soluble granules. During a subsequent dark period, most starch is degraded and converted into sucrose. Sucrose is then exported to storage organs and rapidly growing tissues. The metabolism of starch provides a source of energy in the absence of light.

8. How is CO_2 incorporated into carbohydrate molecules?

The incorporation of CO_2 into carbohydrate molecules is often referred to as the Calvin cycle. The Calvin cycle is a complex series of reactions which can be divided into three phases: carbon fixation, reduction, and regeneration. Many of the Calvin cycle reactions are similar to the pentose phosphate pathway.

9. What is photorespiration and how do some plants avoid it?

Photorespiration is a light-dependent process in which oxygen is consumed and CO_2 is liberated by plant cells which are actively engaged in photosynthesis. Plants use C4 metabolism and crassulacean acid metabolism (CAM) to counteract the photorespiration process.

The C4 metabolism is a mechanism for assimilating CO_2. When the plant opens its stomata at night, CO_2 enters and is incorporated into oxaloacetate. When light is available, photosynthesis provides adequate amounts of ATP and NADPH and CO_2 is released in the bundle sheath cells and converted into sugar. Because the concentration of CO_2 within bundle sheath cells is significantly higher than that of O_2, photorespiration is drastically reduced.

Crassulacean acid metabolism occurs when plant stomata are open and CO_2 is incorporated into oxaloacetate by PEP carboxylase. After malate is formed, it is stored within the vacuole until photosynthesis begins again (i.e., when light is present) and CO_2 is regenerated.

10.   What is the key regulatory enzyme in photosynthesis and how is its activity controlled?

The key regulatory enzyme in photosynthesis is ribulose-1,5-bisphosphate carboxylase-oxygenase (RuBisCO). The activity of RuBisCO is affected by several metabolic signals. In addition to pH and Mg^+, the enzyme is activated by high concentrations of NADPH and fructose-6-phosphate. The enzyme is actively inhibited by fructose-1,6-bisphosphate, O_2, gluconate-6-phosphate, and carboxyarabinitol-1-phosphate.

General Principles

To repeat, think function!

Photosynthesis needs to:
   1. Produce chemical energy (ATP)
   2. Produce reducing equivalents (NADPH)
   3. Fix CO_2 (convert CO_2 into sugars)

Light is used to drive the first two functions; the products of these light-driven reactions (ATP & NADPH) are used to complete the third function.

Function 1: Produce chemical energy (ATP)

This is accomplished in photosystem II. Light energy generates an electron with a large potential energy. These electrons are transferred to a series of electron transport carriers in an analogous manner as in the mitochondrial electron transport chain. As the electrons are transferred, protons are pumped into the thylakoid lumen. Through a process known as photophosphorylation, ATP is synthesized using the proton gradient.

Reactions of photosystem II:

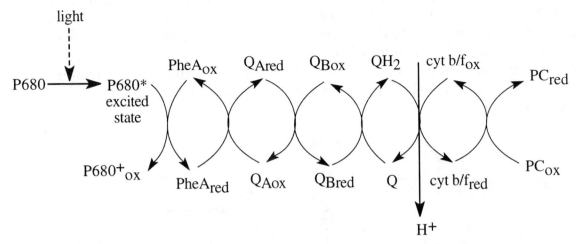

Note: The reduced state means that the carrier has the electrons.

What happens to the reduced plastocyanin (PC)? And what happens to the oxidized P680?

Two possibilities:

1. The electrons from reduced PC can be transferred to oxidized P680. This process is referred to as cyclic photophosphorylation since the electrons are cycled back to the chlorophyll.

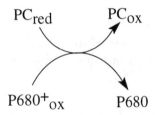

2. The electrons from reduced PC can also be transferred to the chlorophyll in photosystem I, and ultimately used for NADPH production. But there is still a need to reduce P680$^+$. These electrons are replaced when water is oxidized.

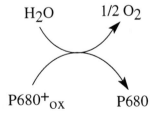

Function 2: Produce reducing equivalents (NADPH)

This is accomplished by the reactions of photosystem I. As mentioned earlier, the electrons from PC_{red} can be transferred to P700 in photosystem I. Another photon of light will excite this electron to a high energy state, after which it will be transferred to a series of electron carrier proteins, starting with chlorophyll and ending with ferredoxin. Ferredoxin ultimately transfers these electrons to $NADP^+$ by way of the enzyme ferredoxin NADP oxidoreductase.

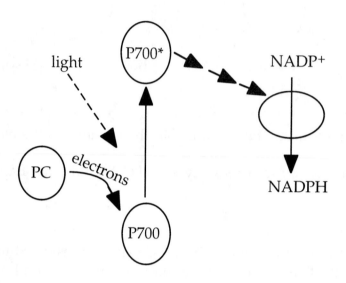

Summary of light reactions

Light is used in photosystem II to promote electrons to an excited state (high potential energy). The electrons are transferred down an energy gradient in order that a proton gradient is generated. The proton gradient is then used for ATP synthesis.

The electrons from photosystem II can then be transferred to photosystem I. More light is needed to excite the electrons to an energy state sufficient for the reduction of $NADP^+$ to NADPH.

Water serves as a source of electrons when electrons are transferred from photosystem II to photosystem I to $NADP^+$. During this process, oxygen is produced from water.

Function 3: Carbon fixation

CO_2 is incorporated and reduced to form intermediates of glycolysis and the pentose phosphate pathway. This process uses three ATPs and two NADPHs for every molecule of CO_2. This series of reactions is called Calvin cycle.

Carbon fixation is accomplished by RuBisCO (Ribulose Bisphosphate Carboxylase-Oxygenase).

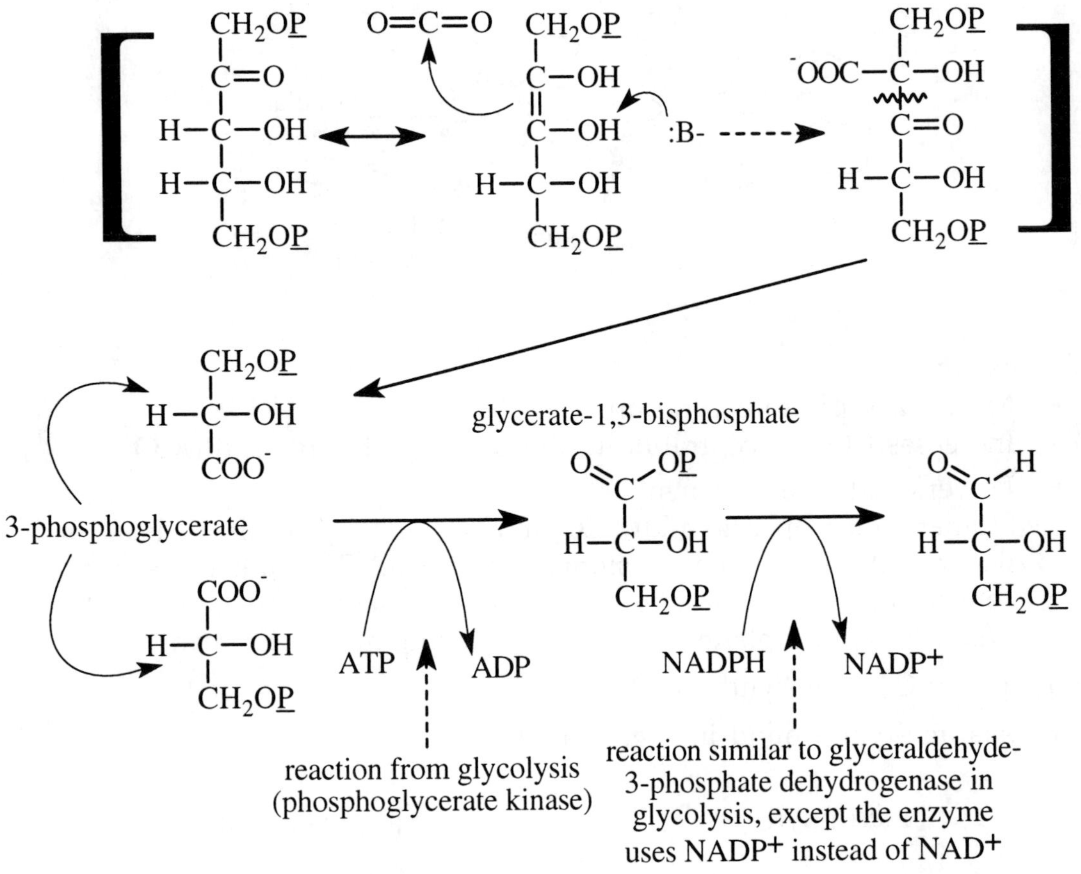

glycerate-1,3-bisphosphate

3-phosphoglycerate

ATP   ADP   NADPH   NADP+

reaction from glycolysis
(phosphoglycerate kinase)

reaction similar to glyceraldehyde-3-phosphate dehydrogenase in glycolysis, except the enzyme uses NADP+ instead of NAD+

The next series of reactions in carbon fixation serve to recycle some of the carbons to form ribulose-1,6-bisphosphate, resulting in the incorporation of another molecule of CO_2. This involves the transaldolase and transketolase reactions we learned about in Chapter 11 on pentose phosphate pathway.

Specialization of Photosynthesis:

Certain plants, called C4 plants, have evolved mechanisms which: 1) reduce water loss, and 2) reduce photorespiration. In C4 plants, the light reactions occur in the mesophyll cells, and the Calvin cycle reactions (CO_2 fixation) occur in the bundle sheath cells. This division of labor allows C4 plants to concentrate CO_2 in the bundle sheath cells at the expense of ATP hydrolysis.

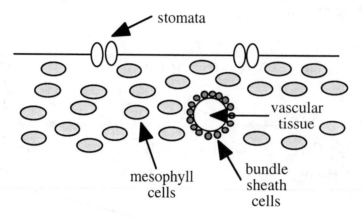

Advantages

- Minimizes photorespiration
  Increases CO_2 concentration in the cells which have RuBisCO

- Lowers water requirement
  Stomata must open to allow CO_2 to enter and O_2 to exit the leaf. But water is also lost when the stomata are open. C4 plants can effectively concentrate CO_2, so that the stomata do not need to be open as often. Water loss in C4 plants is only 10-30% compared to C3 plants.

Two types of C4 photosynthesis:

1. Alanine-type (found in sugar cane)

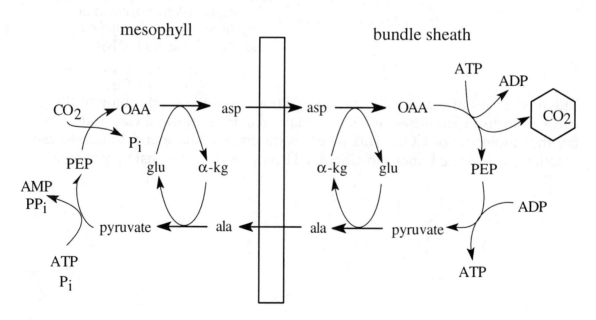

2. Malate-type (found in corn)

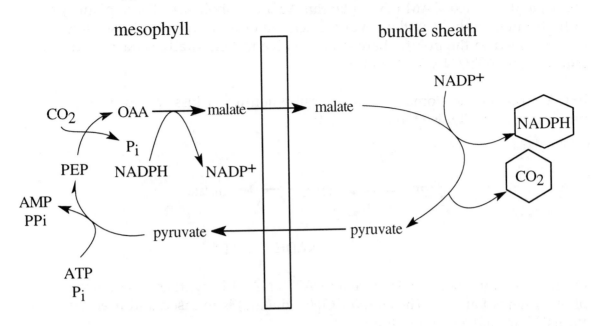

In malate-type C4 plants, both CO_2 and NADPH are transported to the bundle sheath cells.

Bundle sheath cells
- have RuBisCO
- can synthesize ATP (photochemically)
- cannot make NADPH (photochemically)

Mesophyll cells
- do not have RuBisCO
- can synthesize ATP
- can synthesize NADPH

Since the bundle sheath cells cannot synthesize NADPH, they do not make O_2 from H_2O. This further reduces photorespiration.

Another type of photosynthetic specialization is used by desert plants such as cactus, and is called **CAM** (Crassulacean Acid Metabolism). These plants grow in high intensity light without very much water and, in order to survive, have evolved to the point where they are able to temporally separate carbon fixation and ATP/NADPH synthesis.

In CAM plants, the stomata open at night when water loss would be at a minimum, and CO_2 is stored as malate.

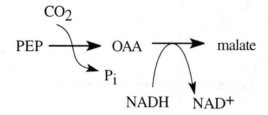

During the day, the stomata close and ATP/NADPH synthesis occurs via photosystems I and II. The stored CO_2 (as malate) is released and used by RuBisCO to make carbohydrates.

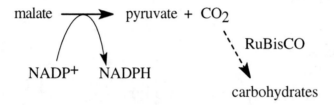

The growth of CAM plants is slowed because the extent of photosynthesis is limited by the amount of CO_2 which is stored as malate during the night.

13   Nitrogen Metabolism I

Objectives

1. What is nitrogen fixation and why is it an important process in the biosphere?

    Nitrogen fixation is the reduction of the atmospheric gas dinitrogen (N_2) to NH_4^+. Nitrogen fixation allows many plants and animals to incorporate nitrogen into the synthesis of many nitrogen-containing biomolecules such as proteins and nucleic acids.

2. What are non-essential and essential amino acids?

    Non-essential amino acids are those amino acids which can be readily synthesized from available metabolites. Essential amino acids are those amino acids which must be provided in the diet.

3. What is the amino acid pool?

    The amino acid molecules which are immediately available for utilization in metabolic processes are referred to as the amino acid pool. Input into the pool consists of amino acids derived from the breakdown of both dietary and tissue proteins. Output takes the form of the excreted nitrogenous end products such as urea and uric acid.

4. What role do transamination reactions play in amino acid metabolism?

    Once amino acid molecules enter cells, the amino groups are available for a variety of synthetic reactions. Transamination reactions allow for metabolic flexibility in which amino groups can be transferred from one α-amino acid to some other α-keto acid. As a result, new amino acids can be synthesized from readily available amino acids. And because transamination reactions are readily reversible, they play an important role in both the synthesis and degradation of the amino acids.

5. What are the amino acid families? Why are amino acids classified into these families?

    The amino acids can be grouped into six families:

<u>Glutamate Family:</u>	glutamate, glutamine, proline, and arginine.
<u>Serine Family:</u>	serine, glycine, and cysteine.
<u>Aspartate Family:</u>	aspartate, asparagine, lysine, methionine, and threonine.
<u>Pyruvate Family:</u>	alanine, valine, leucine, and isoleucine.

Aromatic Family: phenylalanine, tyrosine, and tryptophan.

Histidine Family: histidine.

Amino acids are classified into families because they are ultimately derived from a common precursor molecule.

6. What is one-carbon metabolism?

Amino acids are precursors for many physiologically important biomolecules. The synthesis of these molecules often involves the transfer of carbon groups. Because many of these transfers involve one-carbon groups (e.g., methyl, methylene, methenyl, and formyl), the overall process is referred to as one-carbon metabolism.

7. What role do folic acid and S-adenosylmethionine play in one-carbon metabolism?

Folic acid and S-adenosylmethionine (SAM) are important carriers of one-carbon groups. The carbon groups carried by folic acid (i.e., methyl, methylene, methenyl, and formyl) are bound to the pteridine ring. SAM is a major methyl group donor in one-carbon metabolism. Formed from methionine and ATP, SAM contains an active methyl thioether group which can be transferred to a variety of acceptor molecules.

8. What are biogenic amines?

Biogenic amines are amino acid derivatives which act as neurotransmitters. Examples of these amines are γ-aminobutyric acid (GABA), catecholamines, histamine, and serotonin.

9. What role do the glutathione-S-transferases play in metabolism?

Glutathione (GSH) contributes to the protection of the cell from environmental toxins. GSH accomplishes this by reacting with a large variety of foreign molecules to form GSH conjugates. The bonding of these substances with GSH, which prepares them for excretion, may be spontaneous or catalyzed by GSH-S-transferases (also known as the ligandins). Before their excretion in urine, GSH conjugates are usually converted to mercapturic acids by a series of reactions initiated by γ-glutamyltranspeptidases.

10. How are nucleotides synthesized?

Nucleotides can be synthesized in *de novo* pathways or in "salvage" pathways.

Purine *de novo* pathways entail many steps:

a. Formation of 5-phospho-α-D-ribosyl-1-pyrophosphate (PRPP) catalyzed by phosphate pyrophosphokinase.

b. Displacement of the pyrophosphate group of PRPP by the amine nitrogen of glutamine is catalyzed by glutamine PRPP amidotransferase to form 5-phospho-β-D-ribosylamine.

c. Phosphoribosylglycinamide synthase catalyzes the formation of an amide bond between the carboxyl group of glycine and the amino group of 5-phospho-β-D-ribosylamine. After eight subsequent reactions the ring of IMP is formed.

d. Further reactions can change IMP into either AMP or GMP.

In purine salvage pathways, purine bases are obtained from the normal turnover of cellular nucleic acids or are reconverted into nucleotides. Many cells have mechanisms which retrieve purine bases. Hypoxanthine-guanine phosphoribosyltransferase (HGPRT) catalyzes nucleotide synthesis utilizing PRPP and either hypoxanthine or guanine.

Pyrimidine nucleotide synthesis begins with the formation of carbamoyl phosphate in an ATP-requiring reaction catalyzed by the cytoplasmic enzyme carbamoyl phosphate synthetase II. The carbamoyl phosphate then reacts with aspartate to form carbamoyl aspartate. The closing of the pyrimidine ring is subsequently catalyzed by dihydroorotase to form dihydroorotate. Dihydroorotate is then oxidized by the catalyst dihydroorotate dehydrogenase to form orotate. After its transfer to the cytoplasm, orotate is converted by orotate pyrophosphoribosyl transferase to orotidine-5'-monophosphate (OMP). Uridine-5'-monophosphate (UMP) is produced when OMP is decarboxylated in a reaction catalyzed by OMP decarboxylase. UMP serves as a precursor for the other pyrimidine nucleotides. Two sequential phosphorylation reactions result in the formation of UMP, which then accepts an amide nitrogen from glutamine to form CTP.

11. How are the biosynthesis pathways for heme and chlorophyll similar and how are they different?

The syntheses of heme and chlorophyll are very similar. Although both pathways begin with different starting material, their paths cross when they each form δ–aminolevulinate (ALA). The pathways are identical from ALA to protoporphyrin IX. What distinguishes the remaining pathway is that the protoporphyrin can either be inserted with an Mg^{2+} to form the Mg-protoporphyrin, or it can be inserted with a Fe^{2+} to form protoheme. At this juncture chlorophyll is formed by the addition of a methyl group to the Mg-protoporphyrin and is fully converted by several light-dependent reactions.

General Principles

Amino acids, nucleic acids, and heme all contain nitrogen. Where does the nitrogen come from?

There are two major nitrogen carriers in mammals:

1. glutamate
2. glutamine

Glutamate is used as an **intra**cellular carrier of nitrogen, while glutamine is used as an **inter**cellular carrier of nitrogen.

Why do we need two different carriers for nitrogen? One hypothesis is that glutamate is also used as a neurotransmitter in the brain. If glutamate levels in the blood are elevated, this could cause the brain to "short circuit." In fact, this is an explanation for hot flashes and headaches which some individuals experience as a result of consuming MSG (monosodium glutamate), an ingredient sometimes found in Asian cuisine.

Glutamate and glutamine are synthesized from the citric acid cycle intermediate α-ketoglutarate.

Glutamate dehydrogenase is the enzyme which catalyzes the reductive amination of α-ketoglutarate to form glutamate.

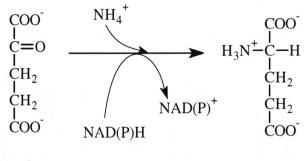

α-ketoglutarate   glutamate

Glutamine synthetase uses glutamate, ATP, and NH_4^+ to form glutamine.

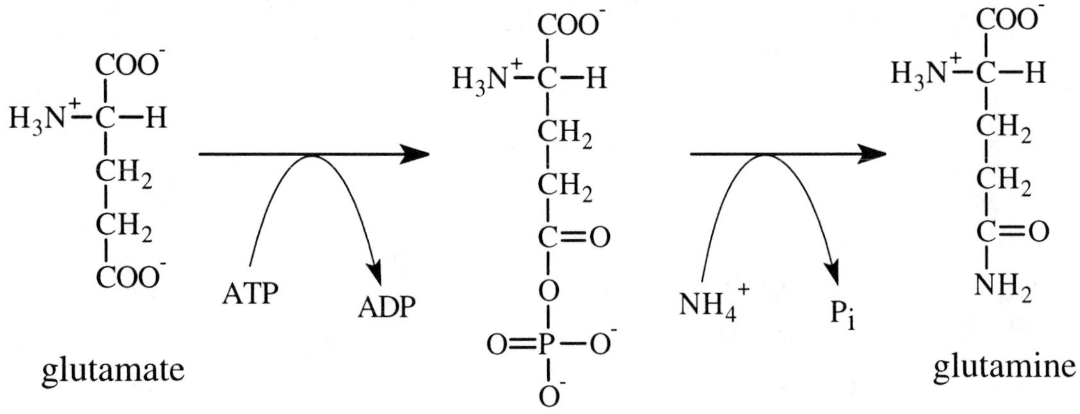

glutamate → (ATP, ADP) → phosphorylated intermediate → (NH_4^+, P_i) → glutamine

One of the challenges of amino acid biosynthesis is constructing the carbon skeletons for all twenty amino acids.

Amino acid biosynthesis is a good example of how biosynthesis is a divergent process. In other words, all twenty amino acids are synthesized from a few intermediates.

Many amino acids are synthesized from the corresponding keto acid.

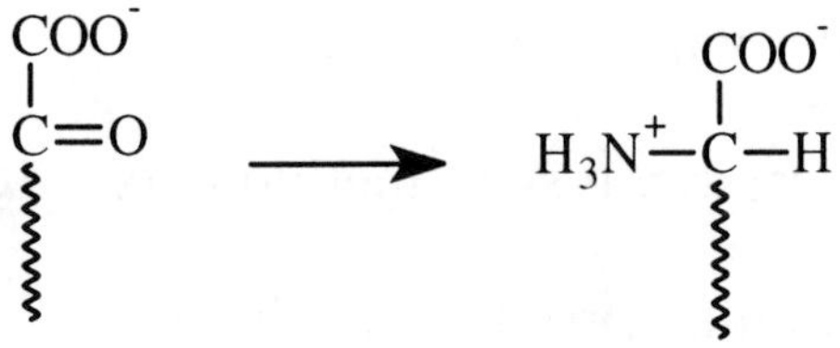

This transformation occurs by **transamination**, in which the nitrogen from glutamate is transferred to the keto acid. This reaction requires pyridoxal phosphate, an important cofactor.

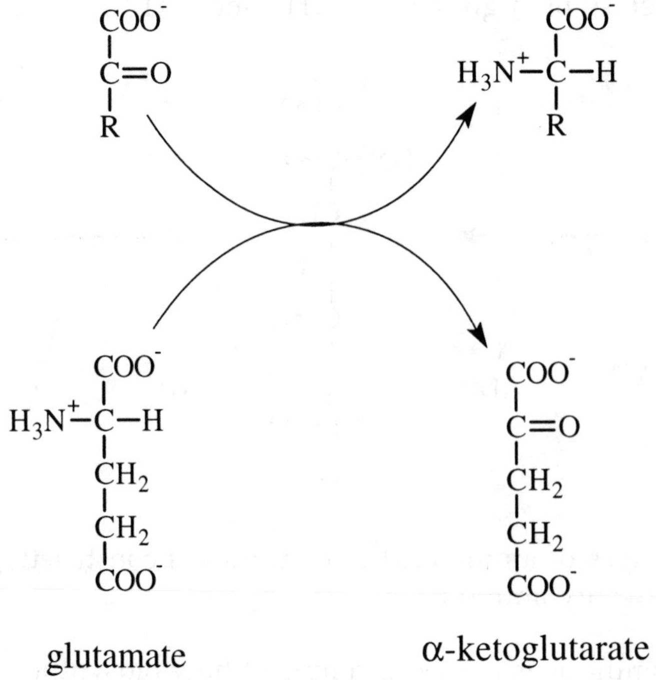

$$\text{glutamate} \rightleftharpoons \alpha\text{-ketoglutarate}$$

glutamate   α-ketoglutarate

One way to study amino acid biosynthesis is to group the amino acids which are synthesized from a common precursor.

There are six basic biosynthetic families:

1. pyruvate →
   alanine, leucine, valine, (isoleucine)
2. oxaloacetate →
   aspartate, asparagine, lysine, methionine, threonine → isoleucine
3. α-ketoglutarate →
   glutamate, glutamine, proline, arginine
4. 3-phosphoglycerate →
   serine, cysteine, glycine
5. phosphoenolpyruvate/erythrose-4-phosphate →
   phenylalanine, tyrosine, tryptophan
6. ribose-5-phosphate →
   histidine

Note: Isoleucine is listed in two families. This is because the carbons in isoleucine are derived from both pyruvate and oxaloacetate.

As an exercise, go through figures 13.5–13.12 in the text to be sure you can follow how a given amino acid is made from the carbon precursor.

Many of the chemical transformations in amino acid biosynthesis are chemically similar to those in glycolysis, citric acid cycle, and fatty acid metabolism, such as hydration, dehydration, decarboxylation, and condensation.

Until now, a reaction we have not studied in this guide is the **one-carbon** transfer reaction used in amino acid biosynthesis, as well as in nucleotide synthesis.

There are two major one-carbon donors:

1. tetrahydrofolate
2. s-adenosyl methionine

Tetrahydrofolate (THF)

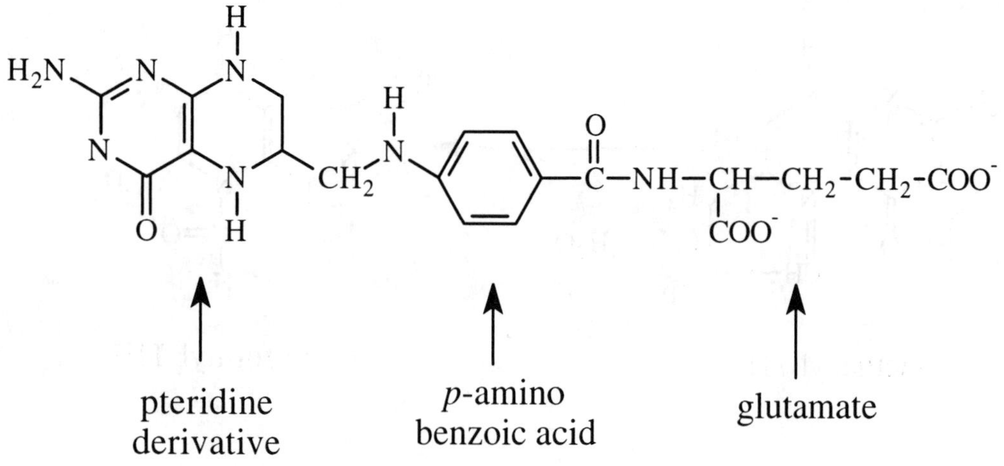
pteridine derivative
p-amino benzoic acid
glutamate

THF can carry carbon in a number of oxidation states.

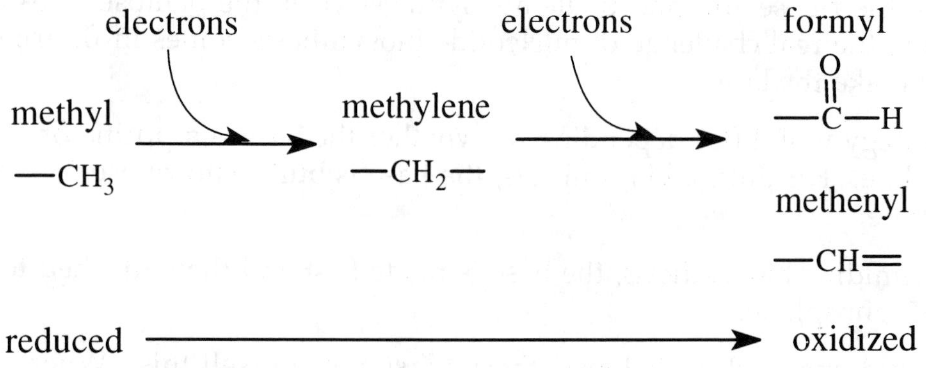
electrons
electrons
methyl
—CH3
methylene
—CH2—
formyl
methenyl
—CH=
reduced
oxidized

The carbon can originate from a number of sources, including glycine, serine, formate, and histidine. Once carbon is attached to THF, the oxidation state can be interconverted.

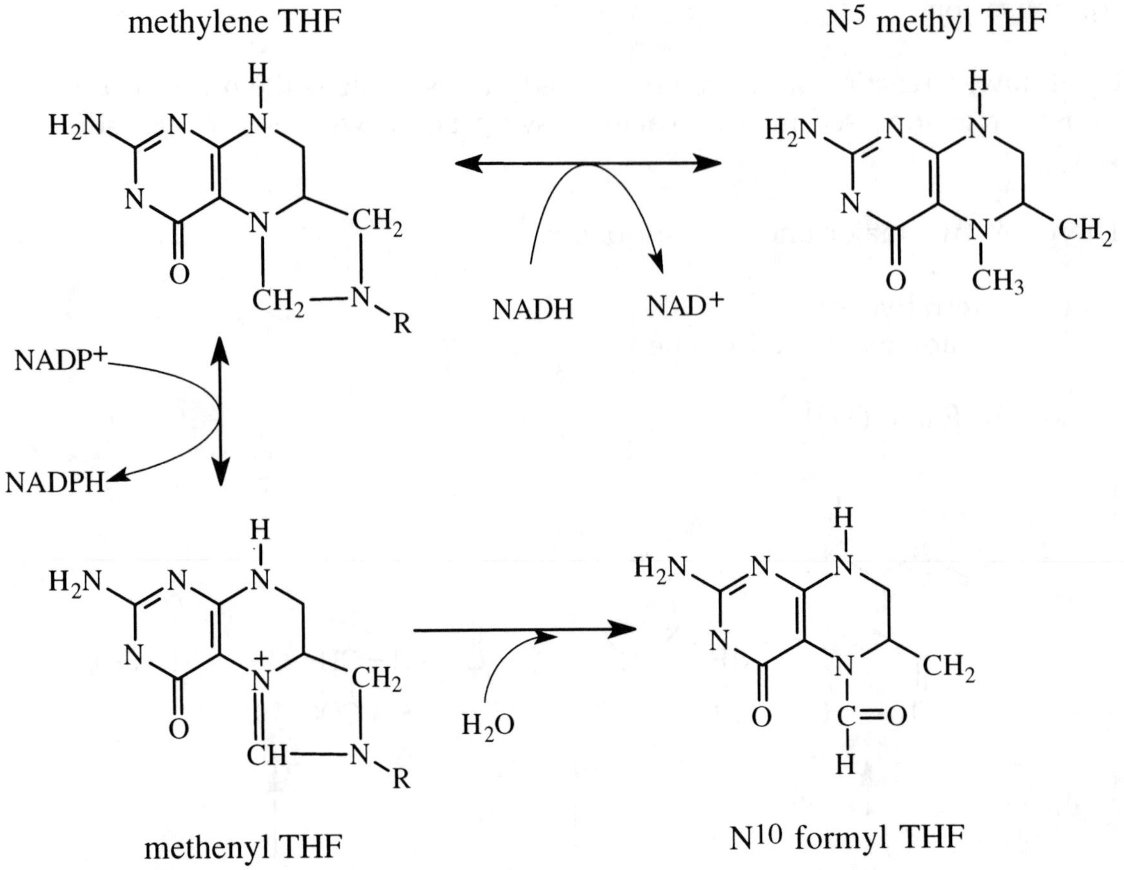
methylene THF
N5 methyl THF
NADH
NAD+
NADP+
NADPH
H2O
methenyl THF
N10 formyl THF

<u>Nucleotide Biosynthesis</u>

Nucleotides consist of a base (purine or pyrimidine), ribose, and phosphate.

Because the ribose and phosphate are synthesized in the pentose phosphate pathway, the real challenge of nucleotide biosynthesis comes in figuring out how to make the base.

The strategy will differ depending on whether the base is a purine or a pyrimidine. For purine biosynthesis, the base is built onto ribose-5-phosphate.

For pyrimidine biosynthesis, the base is made first and then attached to the ribose-5-phosphate.

When studying nucleotide biosynthesis, first ask yourself this: Where do all the atoms come from?

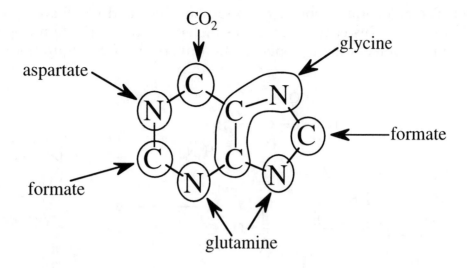

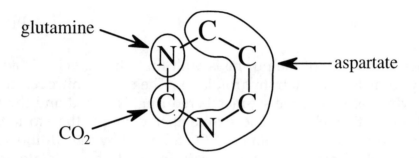

As with many biochemical reactions and pathways, the initial substrate(s) needs to be activated. This is often accomplished by phosphorylation.

For purine biosynthesis, the rings are built onto the ribose. Therefore the ribose-5-phosphate needs to be activated.

α-D-ribose

phosphoribosyl pyrophosphate
(PRPP)

Note: The anomeric carbon of ribose-5-phosphate and PRPP are in α-configuration. This changes in the next step when the amide nitrogen of glutamine displaces the pyrophosphate (inversion of configuration).

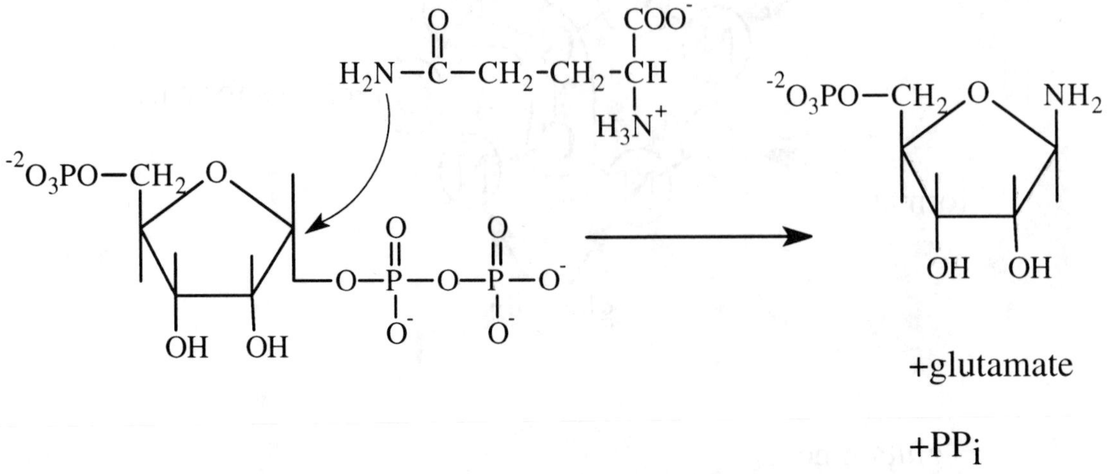

+glutamate

+PP$_i$

The five-membered ring of the purine is built first, then glycine is added followed by formate (from tetrahydrofolate). The second nitrogen from glutamine, which is part of the six-membered ring, is added and the five-membered ring is then closed. The six-membered ring is then formed, starting at the top with the addition of CO_2, followed by the addition of the α-amino nitrogen of aspartate. Formate, again from tetrahydrofolate, is added to the bottom nitrogen (the one from glutamine), and finally the six-membered ring is closed, giving us inosine monophosphate (IMP). IMP can then be converted into AMP or GMP.

The synthesis described above is called *de novo* purine biosynthesis, meaning "from scratch." Purine nucleotides can also be synthesized by salvage pathways. Many nucleic acids, such as RNA, are "turned over," meaning they are synthesized and degraded. Rather than having to synthesize the purine nucleotide *de novo*, the cell can recycle the bases. This makes sense when considering how much energy (ATP equivalents) is expended in purine biosynthesis.

As an exercise, count how many ATP equivalents its takes to synthesize GMP.

The salvage pathways use the "activated" ribose-5-phosphate, PRPP, and the free base.

Adenine + PRPP → AMP + PP$_i$

Guanine + PRPP → GMP + PP$_i$

114

Pyrimidine biosynthesis is much simpler than purine biosynthesis. The nucleotide base is created first and is then attached to the sugar.

The first steps are similar to urea synthesis (Chapter 14).

Carbon dioxide and the amide nitrogen of glutamine are activated at the expense of two ATP equivalents. This forms carbamoyl phosphate, which then condenses with aspartate.

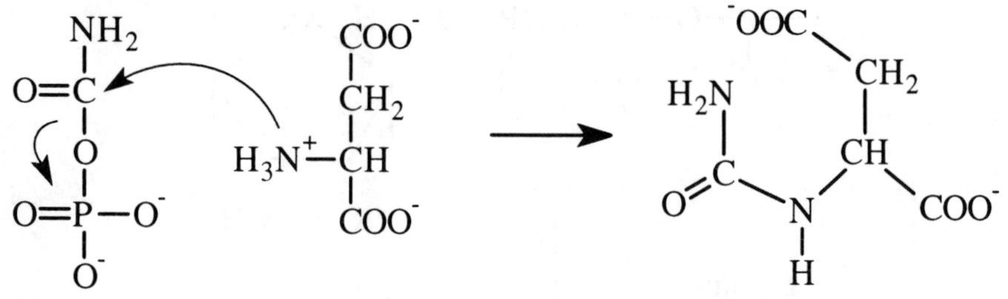

The ring is closed, a double bond is formed, and the ring is attached to an activated ribose-5-phosphate (PRPP). A CO_2 is lost, thereby forming uridine monophosphate (UMP).

UMP is phosphorylated twice by ATP to form UTP.

UMP + ATP $\rightarrow$ UDP + ADP

UDP + ATP $\rightarrow$ UTP + ADP

An amino group, from glutamine, replaces the carboxyl group on UTP to form cytidine triphosphate (CTP).

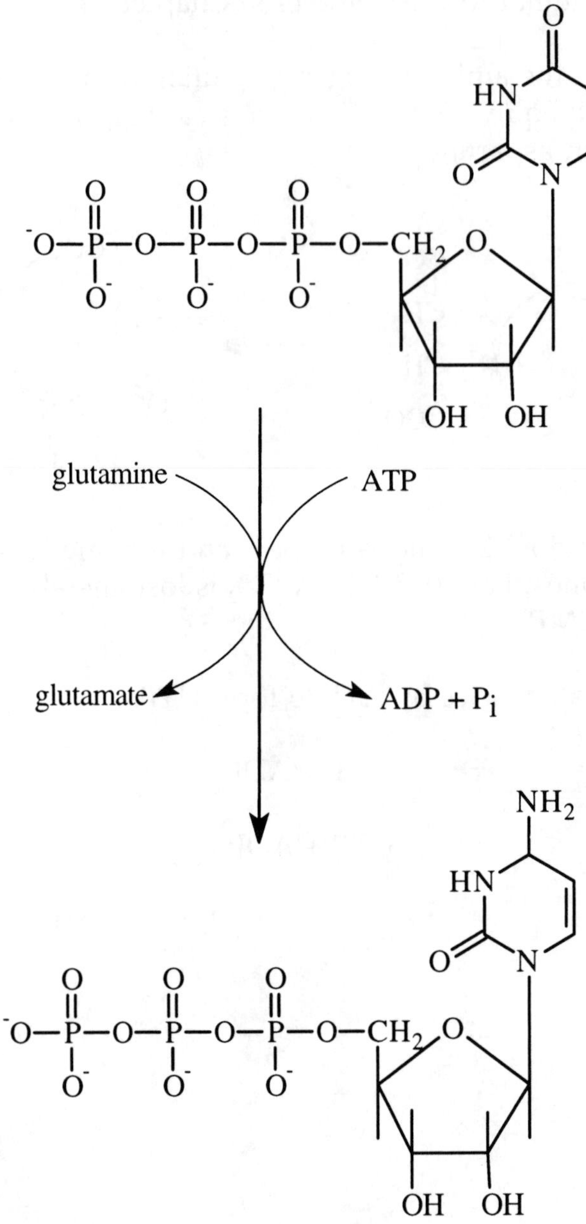

Thus far we've only examined the synthesis of ribonucleotides. How do we get the deoxyribonucleotides?

We must first begin with the nucleotide **di**phosphate.

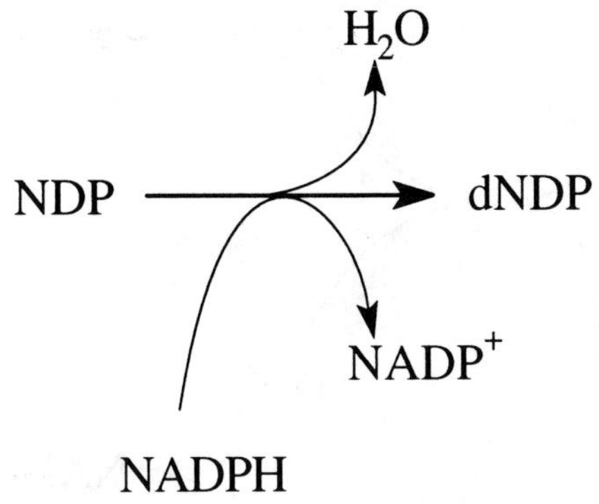

$$NDP \longrightarrow dNDP$$

with H_2O released, $NADPH \rightarrow NADP^+$

The only nucleotide which is missing is deoxythymidine monophosphate (dTMP). dTMP is synthesized from dUMP using the one-carbon donor methylene tetrahydrofolate.

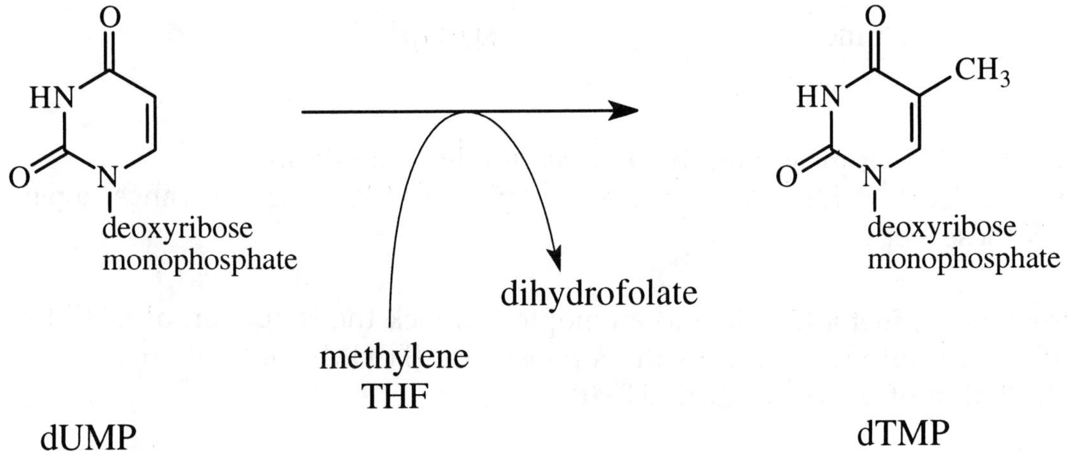

dUMP → dTMP; methylene THF → dihydrofolate

dUMP

dTMP

In the synthesis of dTMP, dihydrofolate (DHF) is produced and needs to be recycled back to tetrahydrofolate as seen below.

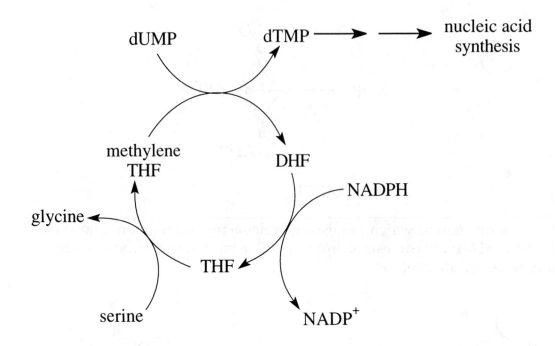

Since cancer cells grow rapidly, they require large amounts of deoxynucleotides for DNA synthesis. Many anti-tumor or anti-cancer agents block these steps.

For example, methotrexate and aminopterin block the reduction of DHF to THF, and fluorodeoxyuridine monophosphate (FdUMP) inhibits the methylation of dUMP to form dTMP.

14   Nitrogen Metabolism II

Objectives

1. How do various animal species dispose of nitrogenous waste?

    Although many variations are observed among species, the following generalizations can be made:

    The nitrogen in amino acids is removed by deamination reactions and converted to ammonia. Many aquatic animals can excrete ammonia, which dissolves in the surrounding water and is quickly diluted. Terrestrial animals must conserve body water. and so they convert ammonia into molecules which can be excreted without a large loss of water. Mammals, for example, convert ammonia to urea. Other animals, such as birds, certain reptiles, and insects, which have more stringent water conservation requirements, are called uricotelic because they convert ammonia to uric acid. In many animals, uric acid is also the nitrogenous waste product of purine nucleotide catabolism.

2. What role does protein turnover play in cellular metabolism?

    The continuous degradation and resynthesis of proteins is a process referred to as protein turnover. Metabolic flexibility is afforded by relatively quick changes in the concentrations of key regulatory enzymes, peptide hormones, and receptor molecules. Protein turnover also protects cells from accumulation of abnormal or damaged proteins. Finally, the process of organismal growth and development are as dependent on timely degradative reactions as they are on synthetic ones.

3. How are proteins targeted for degradation?

    The mechanisms by which proteins are targeted for destruction by ubiquitination or other degradative processes are not fully understood. However, the following features of proteins appear to signal their destruction:

    a. <u>N-terminal residues</u>: The N-terminal residue of a protein is partially responsible for its susceptibility to degradation. For example, proteins with methionine or alanine N-terminal residues have substantially longer half-lives than do those with leucine or lysine.

    b. <u>PEST sequences</u>. Proteins which have extended sequences containing proline, glutamate, serine, and threonine have been observed to possess half-lives of less than two hours.

c. Oxidized residues. Oxidized amino acid residues (i.e., residues which are altered by oxidases or attacked by ROS) promote protein degradation.

4. How are amino groups removed from amino acids?

The removal of the α-amino group from amino acids involves two types of biochemical reactions: transamination and oxidative deamination.

5. How is waste nitrogen incorporated into urea?

The incorporation of nitrogen into urea occurs in a multistep procedure called the urea cycle.

a. Urea synthesis begins with the reaction of NH_4^+ and HCO_3^-, which is catalyzed by carbamoyl phosphate synthetase I, to form carbamoyl phosphate.

b. Carbamoyl phosphate subsequently reacts with ornithine, catalyzed by ornithine transcarbamoylase, to form citrulline.

d. Once citrulline is formed, it is transported to the cytoplasm, where it reacts with aspartate to form argininosuccinate.

e. Argininosuccinate lyase subsequently cleaves argininosuccinate to form arginine and fumarate.

f. In the final reaction of the urea cycle, arginase hydrolyzes arginine to urea and ornithine.

6. How are the citric acid cycle and the urea cycle interrelated?

After fumarate is transported into the mitochondrial matrix, it is hydrated to form malate, a component of the citric acid cycle. The oxaloacetate product of the citric acid cycle can subsequently be converted into aspartate. The aspartate can then be used in the urea cycle to form argininosuccinate by reaction with citrulline. The relationship between the urea cycle and the citric acid cycle is often referred to as the Krebs bi-cycle.

7. By what biochemical pathways are the carbon skeletons of the amino acids degraded into common metabolic intermediates?

The degradation of the amino acid carbon skeletons are classified in terms of their converted end products. The amino acids which are converted into acetyl-CoA are: alanine, serine, glycine, cysteine, threonine, lysine, tryptophan, tyrosine, phenylalanine, and leucine. The amino acids which are converted into α-ketoglutarate are: glutamine, glutamate, arginine, proline, and histidine. The amino acids which are converted into succinyl-CoA are: methionine, isoleucine, and valine. And the amino acids forming oxaloacetate are: aspartate and asparagine.

8. What role does the destruction of neurotransmitters play in the functioning of neurons and muscle cells?

Neurotransmitter degradation in a timely and efficient fashion is required for maintaining precise information transfer in the nervous system. Degradation of neurotransmitters essentially results in the inactivation of the signal.

9. How are dietary nucleic acids degraded?

During digestion, nucleic acids are hydrolyzed to oligonucleotides by enzymes called nucleases. Once formed, oligonucleotides are further hydrolyzed by various phosphodiesterases, a process that produces a mixture of mononucleotides. Nucleotidases remove phosphate groups from nucleotides, thereby yielding nucleosides. These latter molecules are hydrolyzed to free bases and sugar by nucleosidases, which are then absorbed.

10. What degradation products result from the catabolism of purine and pyrimidine bases?

Purines are degraded to uric acid, while pyrimidines are degraded to NH_3, CO_2 and β-alanine or β-aminoisobutryric acid.

General Principles

Amino acid catabolism can be divided into three stages:

1. deamination (removal of the amino group)
2. excretion of the amino nitrogen
3. conversion of the carbon skeleton to a common metabolic intermediate

Deamination is almost always the first step in amino acid catabolism, in which the α-amino group is transferred to α-ketoglutarate to form glutamate.

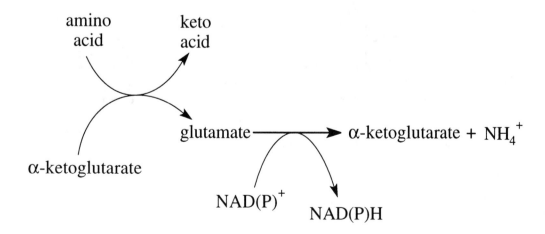

In the liver mitochondrion, glutamate is oxidatively deaminated by glutamate dehydrogenase to release NH_4^+.

Living organisms can excrete nitrogen in 3 ways:

1. ammonium ion
2. urea
3. uric acid

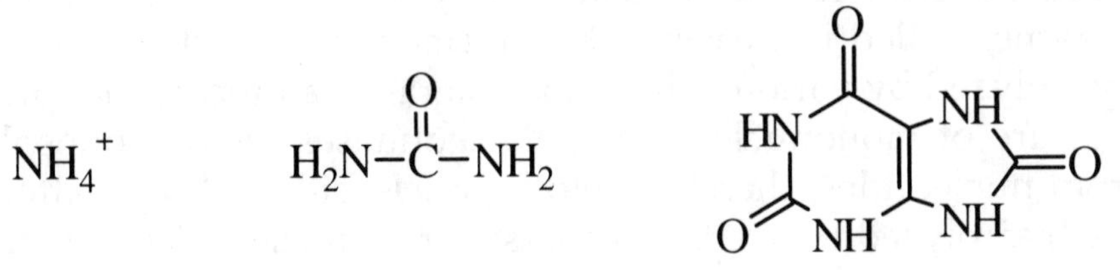

$$NH_4^+$$

ammonium
ion

$$H_2N-\overset{\overset{\displaystyle O}{||}}{C}-NH_2$$

urea

uric acid

The form in which nitrogen is excreted depends on the availability of water. For example, fish excrete nitrogen as ammonia through their gills; mammals excrete urea; birds/reptiles excrete uric acid. This has to do with the solubility of the various nitrogen excretory compounds, and the amount of water necessary and available for excretion.

Urea synthesis was the **first** metabolic cycle discovered and consists of five reactions; two of the reactions occur in the mitochondrion and three in the cytosol.

The overall reaction is as follows:

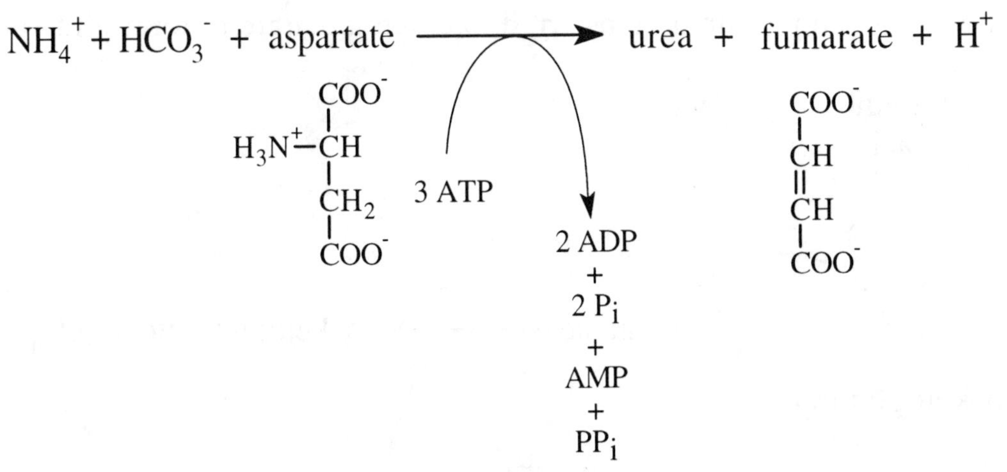

$$NH_4^+ + HCO_3^- + \text{aspartate} \longrightarrow \text{urea} + \text{fumarate} + H^+$$

aspartate structure:
$$H_3N^+ - \overset{\displaystyle COO^-}{\underset{\displaystyle CH_2}{\underset{\displaystyle COO^-}{CH}}}$$

3 ATP

2 ADP
+
2 P$_i$
+
AMP
+
PP$_i$

fumarate structure:
$$\overset{\displaystyle COO^-}{\underset{\displaystyle COO^-}{\underset{\displaystyle CH}{\overset{||}{CH}}}}$$

The first two steps of urea synthesis occur in the mitochondrion. In a reaction similar to what we saw for pyrimidine biosynthesis, at the expense of two ATP equivalents, bicarbonate and ammonium ions form the activated substrate carbamoyl phosphate.

Carbamoyl phosphate then condenses with ornithine to form citrulline. (Note that ornithine is an amino acid; in fact, it looks just like lysine but with one less methylene (-CH$_2$-) group.) Citrulline then exits the mitochondrion via a specific transporter. In the cytosol, the second nitrogen of urea is added from aspartate. This step also requires two ATP equivalents. Argininosuccinate arises from the condensation of aspartate with citrulline. The nitrogen from aspartate is left behind when argininosuccinate is split into arginine and fumarate. The carbons of fumarate come from aspartate. In the final step of urea synthesis, arginine is hydrolyzed to form urea and ornithine.

In examining the overall reaction you probably realized that aspartate and fumarate are an additional reactant and product. An obvious question arises: What happens to the fumarate? The carbons of fumarate can actually recycle to form aspartate via the citric acid cycle. Fumarate is converted into malate and then oxaloacetate. Transamination with glutamate gives rise to aspartate.

Overview of urea synthesis:

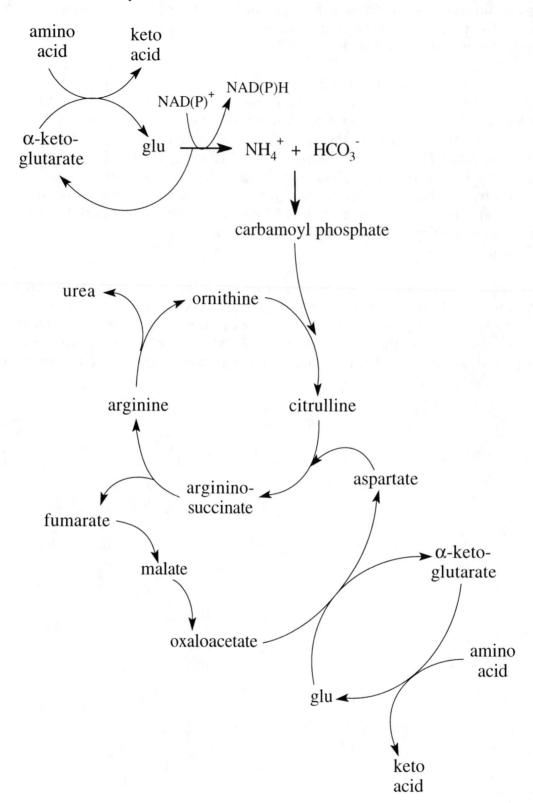

The last stage of amino acid catabolism is the conversion of the carbon skeleton to metabolic intermediates.

The carbon skeletons are eventually degraded to one of seven intermediates. They are:

1.   pyruvate
2.   α-ketoglutarate
3.   succinyl CoA
4.   fumarate
5.   oxaloacetate
6.   acetyl-CoA
7.   acetoacetate

These intermediates can then be catabolized by the citric acid cycle or stored as fatty acids, ketone bodies, or glucose. Amino acids which are degraded to one of the first five intermediates are considered **glucogenic**. This is because these intermediates can enter gluconeogenesis. However, no net synthesis of glucose can occur with acetyl-CoA or acetoacetate. Therefore, amino acids which are degraded to one of these intermediates are considered **ketogenic** since the storage of these carbons results in the formation of ketone bodies or fatty acids.

Nucleotide catabolism

Hint: nucleotide = base + sugar + phosphate
         nucleoside = base + sugar

For both purines and pyrimidines, two steps in degradation are the same. They are:

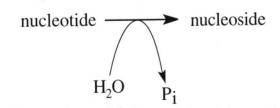

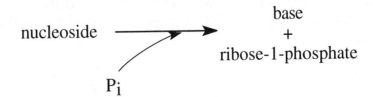

125

Purine bases are degraded to uric acid; pyrimidine bases are converted to malonyl-CoA or methyl malonyl-CoA.

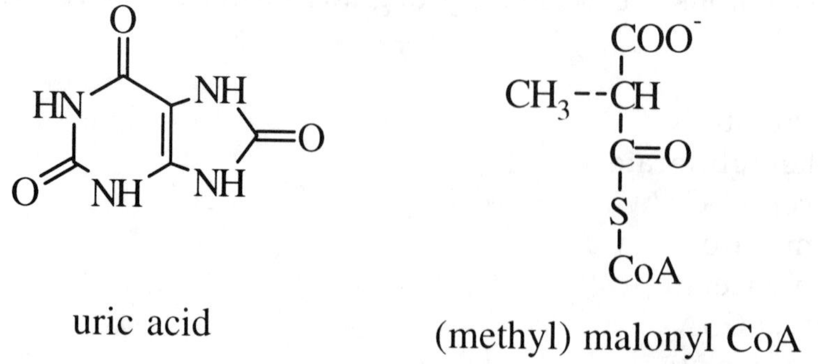

uric acid

(methyl) malonyl CoA

Let's start with purine degradation. For GMP, the phosphate is removed first producing the nucleoside guanosine. The ribose is then removed from the base guanine.

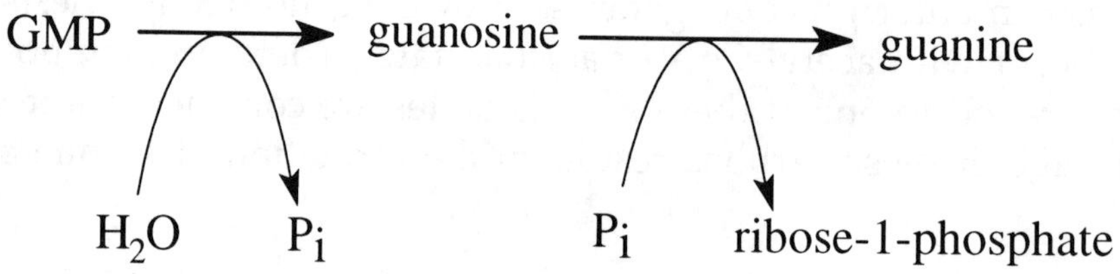

$$GMP \longrightarrow guanosine \longrightarrow guanine$$

$H_2O \qquad P_i \qquad\qquad P_i \qquad ribose\text{-}1\text{-}phosphate$

Guanine loses the amino group, which gives rise to xanthine. Xanthine is then oxidized to form uric acid.

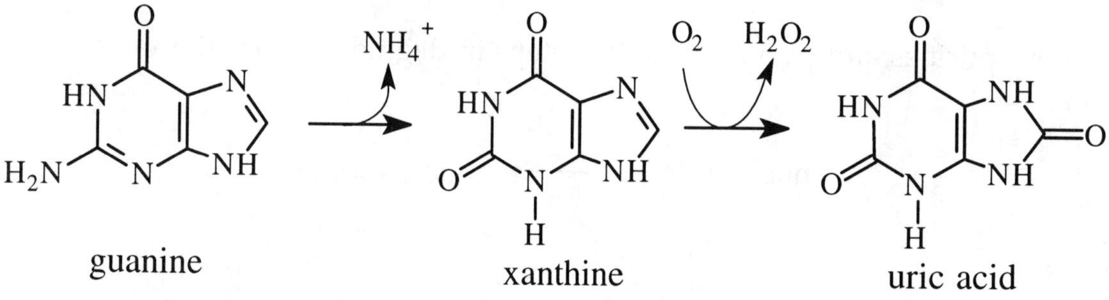

NH_4^+ $\qquad O_2 \quad H_2O_2$

guanine $\qquad\qquad$ xanthine $\qquad\qquad$ uric acid

AMP requires a few additional steps. The phosphate is removed, forming the nucleoside adenosine. The amino group on the base is then removed, producing inosine. The ribose of inosine is removed and the base hypoxanthine is formed. Hypoxanthine is then oxidized to xanthine and later to uric acid.

126

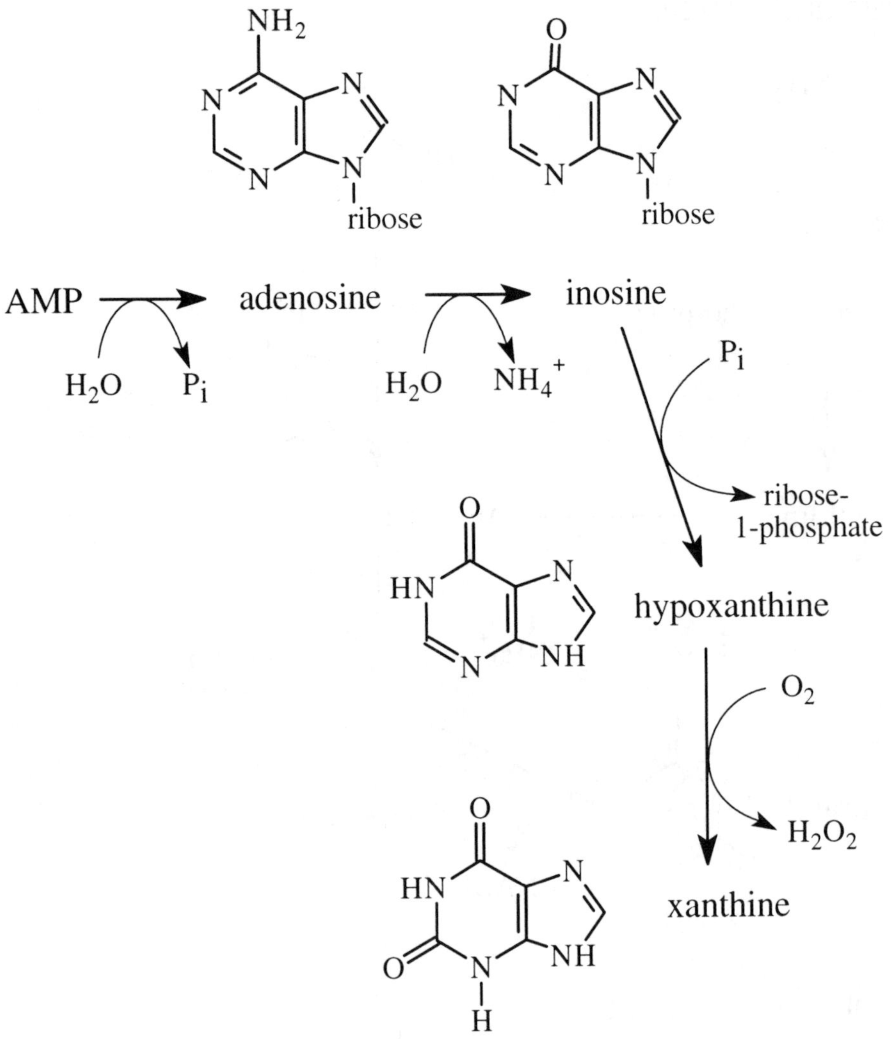

For humans and other primates, uric acid is the final product of purine catabolism. Other animals can degrade uric acid into other nitrogen products such as urea and ammonium ion.

Pyrimidines such as CMP and UMP are degraded to malonyl-CoA. The pyrimidine dTMP, due to its extra methyl group, is degraded to methyl malonyl-CoA.

The phosphate group of CMP is removed, producing the nucleoside cytidine; the amino group is then lost as ammonium ion, thereby forming uridine. Uridine is also formed by removing the phosphate from UMP. The ribose is removed from uridine to form the free base uracil. The double bond in uracil is reduced to form dihydrouracil and the ring is subsequently opened. Once the ring is opened, ammonium and CO_2 are lost producing β-alanine. The last remaining nitrogen is transferred to α-ketoglutarate, and the resulting aldehyde is oxidized with NAD^+ to form the coenzyme A derivative called malonyl-CoA.

Catabolism of Pyrimidines

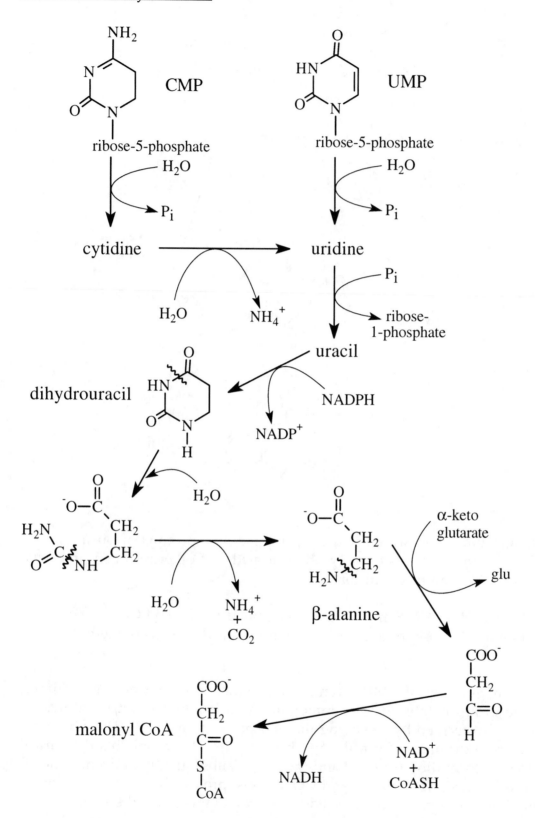

15   Integration of Metabolism

Objectives

1. What are the metabolic contributions of each major organ in
    mammals?

    Small intestine: The most obvious role of the small intestine is aiding
    in the digestion and absorption of nutrients such as carbohydrates,
    lipids, and proteins. Nutrient absorption by the enterocytes of the
    small intestine is an extremely vital and complicated process that
    involves numerous enzymes and transport mechanisms. The small
    intestine also appears to be a major site for glutamine metabolism.
    During the digestive process enterocytes obtain glutamine from dietary
    protein.

    Liver: In addition to its key roles in carbohydrate, lipid, and amino acid
    metabolism, the liver monitors and regulates the chemical
    composition of blood and synthesizes several plasma proteins. Because
    of its metabolic flexibility, the liver is responsible for reducing
    fluctuations in nutrient availability which are caused by drastic dietary
    changes such as intermittent feeding and fasting. The liver also plays a
    critically important protective role in the processing of foreign
    molecules.

    Muscle: The energy sources which are used to provide ATP for muscle
    contraction depend on the degree of muscular activity and the physical
    status of the individual concerned. During fasting and prolonged
    starvation, some skeletal muscle protein is degraded to provide amino
    acids to the liver for gluconeogenesis. The cardiac muscle in the heart
    must continuously contract to sustain blood flow throughout the body.
    To maintain its continuous operation, cardiac muscle relies mainly on
    fatty acids.

    Adipose tissue: The role of adipose tissue is primarily the storage of
    energy in the form of triacylglycerols. Depending on physiological
    conditions, adipocytes convert excess nutrient molecules to fat deposits,
    or degrade fat to generate energy-rich fatty acids and glycerol.

    Brain: The brain is ultimately responsible for directing most metabolic
    processes in the body. Much of the body's hormonal activity is
    controlled either directly or indirectly by the hypothalamus and the
    pituitary gland. Under normal conditions the brain uses glucose as its
    sole fuel. Under conditions of prolonged starvation, the brain can
    adapt by using ketone bodies as an energy source.

<u>Kidney</u>. The kidney has several important functions which include:

a. The filtration of blood plasma, which results in the excretion of water-soluble waste products;

b. Reabsorption of electrolytes, sugars, and amino acids from the filtrate; and

c. The regulation of the body's water content.

Energy is provided largely by fatty acids and glucose. Under normal conditions the small amounts of glucose which are formed by gluconeogenesis are used only within certain kidney cells. The rate of gluconeogenesis increases during starvation and acidosis.

2. Why do animals consume food only intermittently?

The consumption of food on an intermittent basis is possible because of elaborate mechanisms for storing and mobilizing energy-rich molecules derived from food. During the feeding phase, food is consumed, digested, and absorbed. Absorbed nutrients are then transported to various organs where they are either utilized or stored. When a fast is prolonged, several strategies are used to maintain blood glucose levels. For example, fatty acids are mobilized from adipose tissue, and these fatty acids provide an alternative energy source.

3. What is the role of hormones in the feeding-fasting cycle?

Changes in the status of various biochemical pathways which occur during transition between the nutritional phases of feeding and fasting illustrate metabolic integration and the profound regulatory influence of hormones.

4. What specific effects do glucagon and insulin have on metabolism?

During the feeding phase, hormones such as gastrin, secretin, and cholecystokinin contribute to the digestive process. They stimulate the secretion of various enzymes or digestive aids such as bicarbonate and bile. As glucose moves from the small intestine to the liver via the blood, β-cells within the pancreas are stimulated to release insulin. The release of insulin triggers several processes which ensure the storage of nutrients. In addition, insulin also influences amino acid metabolism. For example, insulin promotes the transport of amino acids into cells.

During a fast, as blood glucose and insulin levels return to normal, glucagon release is triggered. Glucagon prevents hypoglycemia by promoting glycogenolysis and gluconeogenesis in liver. Norepinephrine increases mobilization of fatty acids from adipose tissue during the post-absorptive state, and glucagon results in increased gluconeogenesis.

5. What is the hormone cascade system and how is it controlled?

In mammals, most metabolic activities are controlled to a certain extent by hormones. To ensure proper control of metabolism, the synthesis and secretion of many hormones are regulated by a complex cascade mechanism that is ultimately controlled by the central nervous system. Various types of sensory signals are received by the hypothalamus, an area in the brain that links the nervous and endocrine systems. Once it is appropriately stimulated, the hypothalamus induces the secretion of several hormones produced by the anterior lobe of the pituitary gland.

6. How are the structures of the hypothalamus and pituitary gland related to the functioning of the processes they regulate?

The pituitary gland, which is attached to the hypothalamus by the pituitary stalk, consists of two distinct parts: the anterior lobe, or adenohypophysis, and a posterior lobe, or neurohypophysis. The hypothalamus synthesizes a series of specific peptide-releasing hormones. Hormones then pass into a capillary bed referred to as the hypothalamohypophyseal portal system, which transports them directly to the adenohypophysis. These peptides stimulate specific cells to synthesize and secrete one or more types of hormone. The hormones of the anterior pituitary are sometimes referred to as tropic, since they both stimulate the synthesis and release of hormones from other endocrine glands.

The anatomy and function of the posterior pituitary differ from those of the anterior lobe. The hormones secreted by the neurohypophysis are actually synthesized in separate types of neurons which originate in the hypothalamus.

7. What role does the down-regulation of receptors play in metabolism?

The reduction in cell surface receptors in response to stimulation by specific hormone molecules is called down-regulation. In down-regulation, receptors are internalized by endocytosis. Depending on the cell type and several other factors, the receptors may eventually be recycled back to the cell surface or degraded.

8. Which diseases result from the overproduction or underproduction of hormones?

The oversecretion of hormone molecules is most often caused by a tumor. Several types of pituitary tumors cause endocrine disease. Cushing's disease is a condition characterized by obesity, hypertension and elevated blood glucose levels. Gigantism, in which there is a pronounced growth of long bones, is caused by excessive secretion of growth hormone during childhood. In adulthood, excessive growth hormone production causes acromegaly, a condition in which connective tissue proliferation and bone thickening result in coarsened and exaggerated facial features, as well as enlarged hands and feet.

However, not all hypersecretion diseases are caused by tumors. For example, Graves disease, the most common type of hyperthyroidism, is an autoimmune disease.

Inadequate hormone production has a variety of causes. The most common of these are the autoimmune destruction of hormone-producing cells, genetic defects, or an inadequate supply of precursor molecules. Addison's disease is a disorder in which there is inadequate adrenal cortex function. The most common cause of Addison's disease is the autoimmune destruction of the adrenal gland. Hypothyroidism may be the result of autoimmune disease (Hashimoto's disease) or the deficient synthesis of thyroid hormones such as TSH and TRH. Because adequate ingestion of iodine is a prerequisite for thyroid hormone synthesis, iodine deficiency can cause hypothyroidism. In children, a thyroid hormone deficiency called cretinism causes depressed growth and mental retardation. In adults, myxedema, also a thyroid hormone deficiency, results in symptoms such as edema and goiter.

Growth hormone deficiency may be hereditary or a consequence of a pituitary tumor or head trauma. Congenital growth hormone deficiency results in shortened stature known as dwarfism.

Diabetes insipidus is a malady characterized by the passage of copious amounts of very diluted urine. It is caused by either an inadequate synthesis of vasopressin or the failure of the kidney to respond to vasopressin.

9. What are the two major forms of diabetes mellitus?

    In insulin-dependent diabetes mellitus, also called type I diabetes, inadequate amounts of insulin are secreted because of the destruction of the β-cells in the pancreas. Noninsulin-dependent diabetes mellitus, also called type II or adult onset diabetes, is caused by a loss of responsiveness of the target tissue to insulin.

10. What roles do growth factors play in animals?

    Growth factors are a group of hormones, such as polypeptides and proteins, which promote cell growth as well as cell division and differentiation.

11. What are the major second messengers and how do they mediate hormonal messages?

When a hormone molecule binds to a plasma membrane receptor, an intracellular signal called a second messenger is generated. It is the second messenger that actually delivers the hormonal message inside the cell. Signal transduction serves to amplify the original signal. The major second messengers are cAMP, cGMP, DAG, IP_3, and Ca^{2+}.

12. What are the major plant hormones?

The important hormones which have so far been investigated include auxins, gibberellins, cytokinins, abscisic acid, and ethylene.

Additional Concepts

Macro-catabolism occurs when large macromolecules are broken down into their individual components. For mammals, this process is commonly referred to as digestion and occurs in the mouth, stomach, and intestines.

$$\text{proteins} \xrightarrow{\text{proteases}} \text{amino acids}$$

$$\text{fats} \xrightarrow{\text{lipases}} \text{fatty acids}$$

$$\text{starch} \xrightarrow{\text{amylases}} \text{sugars}$$

These individual components then feed into the central catabolic pathways such as glycolysis and the citric acid cycle.

Nutrition and catabolism

The energy content of food is measured as calories. Calories are determine by the amount of heat given off when a food substance is completely oxidized. The caloric value of food is slightly reduced in our bodies due to incomplete digestion and metabolism. The general caloric content of various foods is:

protein	four calories/gram
fat	nine calories/gram
carbohydrates	four calories/gram
alcohol	seven calories/gram

What factors are involved in the expenditure of calories?

There are four principal factors to consider:

Surface area (related to height and weight): This is related to the amount of heat loss from the body. A lean individual has a greater surface area and therefore a greater energy requirement.

Age: The basal metabolic rate (BMR) is a measure of energy utilization at rest. BMR is related to growth and lean muscle mass. Infants and children are rapidly growing and thus have a higher BMR.

Sex: Females tend to have a lower BMR than males. This generalization arises from observations that women have a smaller percentage of lean muscle mass and that the female sex hormones may also affect metabolism.

Activity levels (or exercise): Long-term effects are more important than the actual calories "burned" during exercise. Regular exercise can increase the BMR and therefore increase the overall calories expended. A general exercise program designed to increase lean muscle mass will help increase the BMR.

Diet-induced thermogenesis

Most Americans consume an excess of calories than are needed for immediate energy production. We all know someone who can eat just about anything and never get fat. So what happens to those excess calories?

Excess calories have two major fates. The first is storage, usually as fat. Excess calories can also be converted to heat, a process called **thermogenesis**. The degree of thermogenesis is influenced by a number of factors, including genetics, diet, and exercise.

The biochemical mechanism for thermogenesis is unknown. In rodents, such as rats and mice, a specialized fat tissue called brown adipose tissue is believed to be involved in thermogenesis. In humans, skeletal muscle appears to be responsible for heat generation and dissipation. One proposed mechanism for thermogenesis involves the shuttle systems for transferring electrons from the cytosol to the mitochondrion. The glycerol phosphate shuttle transfers electrons from cytosolic NADH to mitochondrial $FADH_2$. This gives rise to approximately two ATP equivalents. The other shuttle is the malate-aspartate shuttle in which electrons are moved from cytosolic NADH to mitochondrial NADH, thereby giving rise to approximately three ATP equivalents. Thus when the glycerol phosphate shuttle is used, one ATP equivalent of energy is lost. The energy has to go somewhere; the belief is that it is dissipated as heat. This hypothesis is consistent with the observation that muscles use the glycerol phosphate shuttle to a greater extent than other tissues.

16   Nucleic Acids

Objectives

1.   What is the central dogma?

   The central dogma describes the flow of genetic information from DNA through RNA and eventually to proteins. The central dogma proceeds from replication of DNA, transcription of RNA, and translation of protein.

2.   How is DNA damaged?

   Mutations take many forms, from point mutations to gross chromosomal abnormalities. They are caused by a variety of factors which include the chemical properties of the bases themselves and various chemical processes, as well as the effects of xenobiotics, radiation, and viruses.

   Although DNA is a relatively stable molecule, it is vulnerable to certain types of structural change. Tautomeric shifts are spontaneous changes in nucleotide base structure which consist of the interconversion of amino and imino groups as well as keto and enol groups. Base mispairing and transition mutation, in which a pyrimidine base is substituted for another pyrimidine, or a purine is substituted for another purine, may occur as a result.

   A large number of xenobiotics can potentially damage DNA. The more important of these molecules belong to the following classes:

   a. Base analogs: Because their structures are similar to normal nucleotide bases, base analogs can be inadvertently incorporated into DNA.

   b. Alkylating agents: Alkylation is a process in which electrophilic substances attack nucleophilic atoms.

   c. Nonalkylating agents: Once consumed, these molecules can be converted to highly reactive derivatives by biotransformation reactions. Damage occurs primarily because this type of chemical modification results in the prevention of base pairing.

   d. Intercalating agents: Certain planar molecules can distort DNA because they insert themselves between stacked base pairs of the double helix.

   e. Radiation: Various forms of radiation (e.g., UV, X-rays, and γ-rays) can alter DNA structure. Examples of changes induced by radiation include bond breakage and ring opening.

3.   What historical developments led to the discovery of DNA structure?

1865   The scientific revolution that eventually necessitated the determination of DNA structure began when Gregor Mendel discovered the basic rules of inheritance.

1869   Friedrich Miescher discovered "nuclein," later renamed nucleic acid.

1882-1897 The chemical composition of DNA was determined largely by Albrecht Kossel.

1928   Fred Griffith proposed the concept of transmission of genetic information between bacterial cells.

1944   Avery and McCarty demonstrated that the digestion of DNA by deoxyribonuclease inactivated the transforming agent, concluding that genetic information was carried by DNA.

1952   By using T2 bacteriophage, Hershey and Chase demonstrated the separate functions of viral nucleic acid and protein. These experiments reconfirmed DNA as the genetic material.

4.   What information did Watson and Crick use to determine DNA structure?

By the early part of the decade, the chemical structures and dimensions of the nucleotides had been elucidated, and it was known that adenine:thymine and guanine:cytosine existed as 1:1 ratios in DNA. X-ray diffraction studies performed by Rosalind Franklin indicated that DNA was a symmetrical molecule, in all likelihood a helix; the diameter and pitch of the helix was estimated by Wilkins and Stokes.

5.   What are the structural differences among A-DNA, B-DNA, H-DNA, and Z-DNA?

B-DNA: Base pairs are at right angles; 10.4 base pairs per helical turn; each helical turn occurs at 3.4 nm; and the diameter is 2.0 nm.

A-DNA: Base pairs tilt twenty degrees away from the horizontal; eleven base pairs per helical turn; each helical turn occurs at 2.8 nm; and the diameter is 2.6 nm.

Z-DNA: Base pairs are in a zigzag conformation; twelve base pairs per helical left turn; each helical turn occurs at 4.5 nm; and the diameter is 1.8 nm.

H-DNA: Under certain circumstances, a long DNA segment consisting of a polypurine strand hydrogen-bonded to a polypyrimidine strand can form a triple helix. The formation of the triple helix, also referred to as H-DNA, depends on the formation of non-conventional base pairs, which occurs without disrupting the Watson-Crick base pairs.

6. What is the structural significance of DNA supercoiling?

DNA supercoiling is now known to facilitate several biological processes. Examples of this include the packaging of DNA into a compact form, as well as DNA replication and transcription.

7. What are nucleosomes?

Nucleosomes are a repeating structural element in eukaryotic chromosomes, composed of a core of eight histone molecules with about 140 base pairs of DNA wrapped around the histone. An additional sixty base pairs connect adjacent nucleosomes.

8. What are the three most prominent forms of RNA?

The three most prominent forms of RNA are ribosomal RNA, transfer RNA, and messenger RNA. Ribosomal RNA is the most abundant form of RNA and is responsible for the formation of ribosomes. Transfer RNA molecules transport individual amino acids to ribosomes, where they are properly aligned during ongoing protein synthesis. Messenger RNA molecules contain the information required to direct the synthesis of specific polypeptides.

9. What properties of viruses make them useful research tools for biochemists?

Driven in large part by the role viruses play in numerous diseases, viral research has been of tremendous benefit to biochemistry. Because a virus essentially subverts normal cell function to produce more viruses, a viral infection can provide unique insight into cellular metabolism. Several eukaryotic genetic mechanisms have been elucidated with the aid of viruses and/or viral enzymes. Viral research has also provided substantial information concerning genome structure and carcinogenesis. Viruses have also been invaluable in the development of recombinant DNA technology.

10. What types of genomes can viruses possess?

Despite the fact that viruses are acellular and cannot carry out metabolic activities on their own, viruses can wreak havoc on living organisms. Each type of virus infects a specific type of host (or small set of hosts). Although most viruses possess double-stranded DNA or single-stranded RNA, examples of single-stranded DNA and double-stranded RNA genomes have been observed.

11. What is the difference between lytic and lysogenic viruses?

Lytic viruses destroy the host cell; lysogenic viruses integrate their genome into the host cell's genome.

12. After many years of research and expending billions of dollars, why is AIDS still considered an incurable disease?

HIV is a retrovirus that is believed to cause AIDS. Retroviruses are a class of RNA viruses which possess a reverse transcriptase activity that converts their RNA genome to a DNA molecule. This DNA is then inserted into the host cell genome, causing a permanent infection. Because the viral genome mutates frequently (i.e., its surface antigens become altered), an AIDS vaccine has been difficult to develop.

General Principles

All the information necessary to construct a replica of an organism resides in the genetic material or genome. The genetic material of most organisms is composed of DNA; however, some viruses use RNA.

Nucleic acids are polymers of nucleo**tides**. A nucleotide is a sugar, base, and phosphate, while a nucleo**side** contains only the sugar and base.

The sugar in nucleotides can either be ribose (for RNA) or deoxyribose (for DNA). The atoms of the sugar are numbered from the anomeric carbon and are distinguished from the base atoms by the prime (') notation.

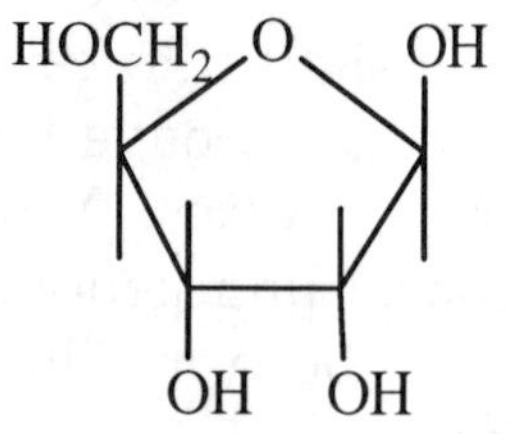

ribose
β-D-ribofuranose

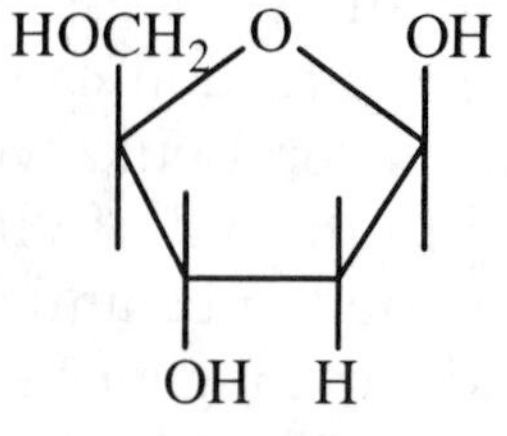

dexoxyribose
2-deoxy β-D-ribofuranose

The base of nucleotides are classified as either pyrimidines or purines, and the atoms are numbered as indicated below.

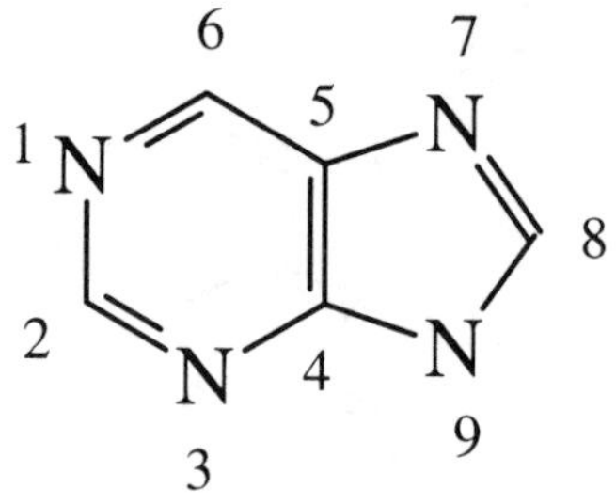

purine 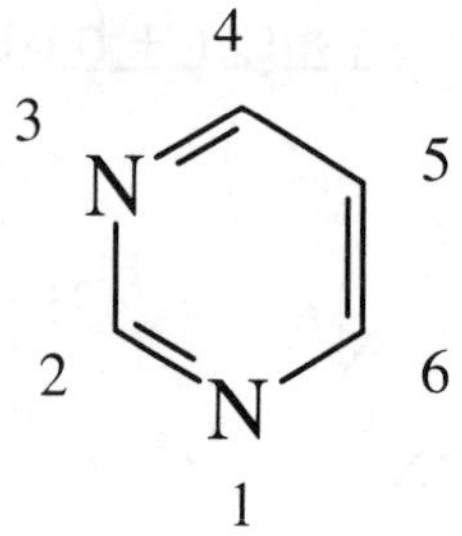   pyrimidine

The following is a list of the common bases, nucleosides, and nucleotides found in RNA and DNA.

Purine Bases

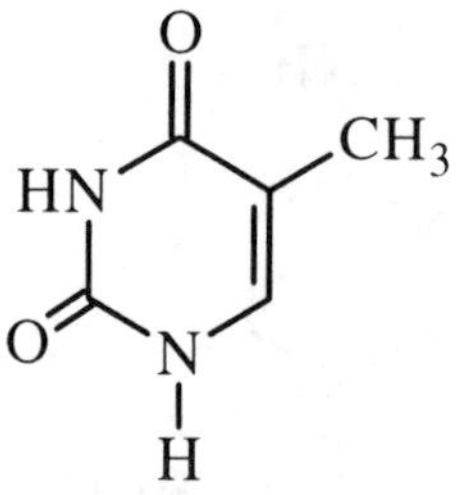

thymine 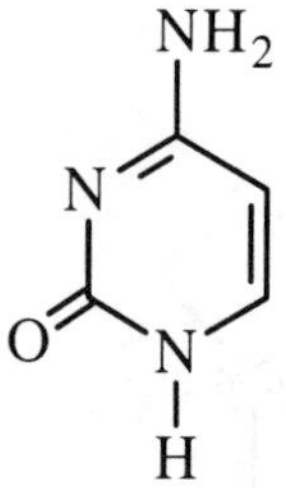   cytosine 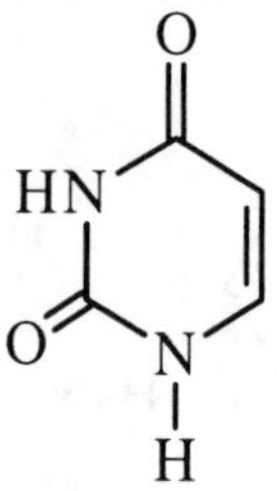   uracil

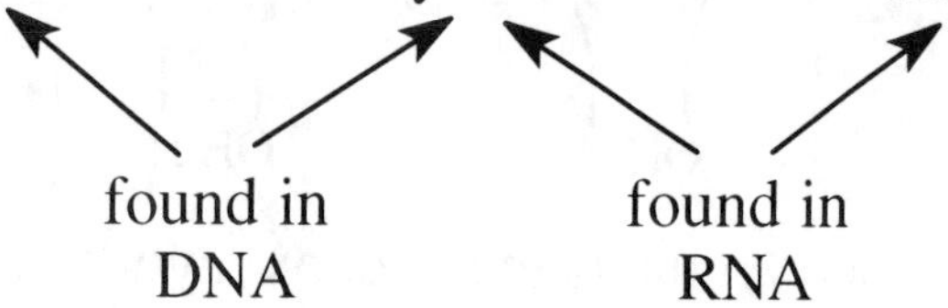

found in   found in
DNA   RNA

Pyrimidines Bases (found in both RNA and DNA)

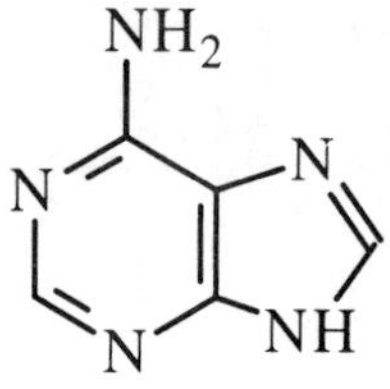

adenine 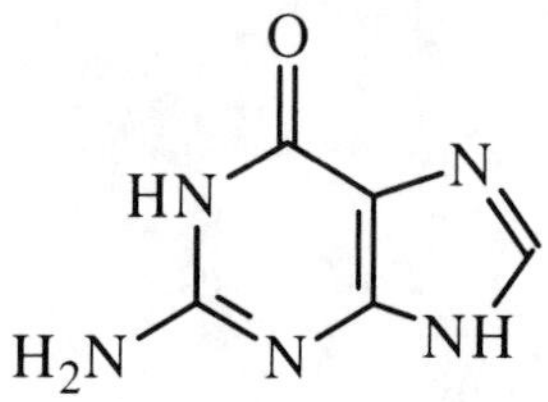   guanine

139

Nucleosides (sugar + base)

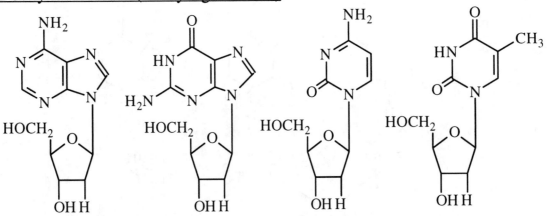

adenosine   guanosine   cytidine   uridine

Deoxynucleosides (deoxysugar + base)

deoxyadenosine   deoxyguanosine   deoxycytidine   deoxythymidine

140

Nucleotides (sugar + base + phosphate)

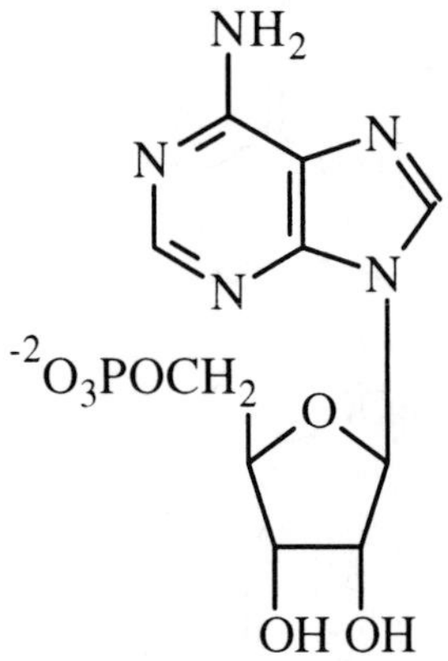

adenosine
5'-monophosphate
(AMP)

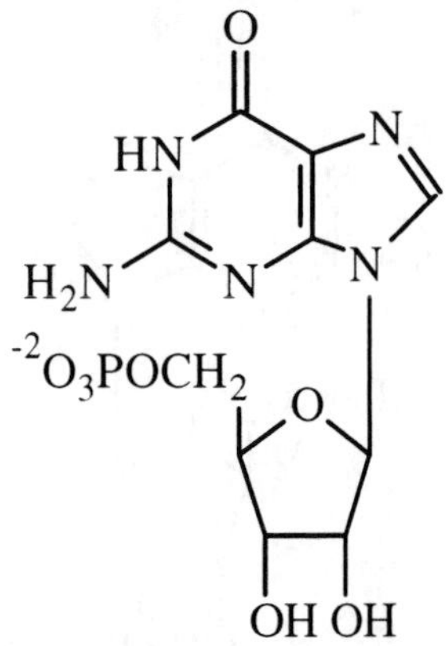

guanosine
5'-monophosphate
(GMP)

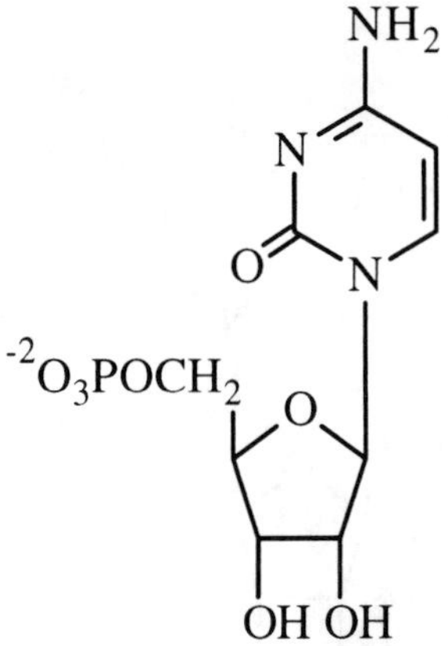

cytidine
5'-monophosphate
(CMP)

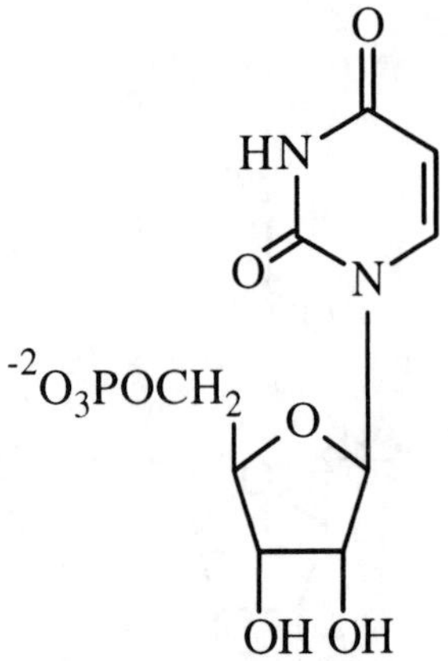

uridine
5'-monophosphate
(UMP)

Deoxynucleotides (deoxysugar + base + phosphate)

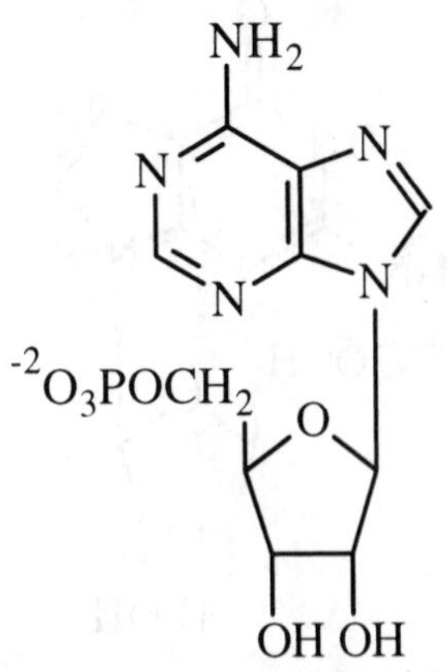

deoxyadenosine
5'-monophosphate
(dAMP)

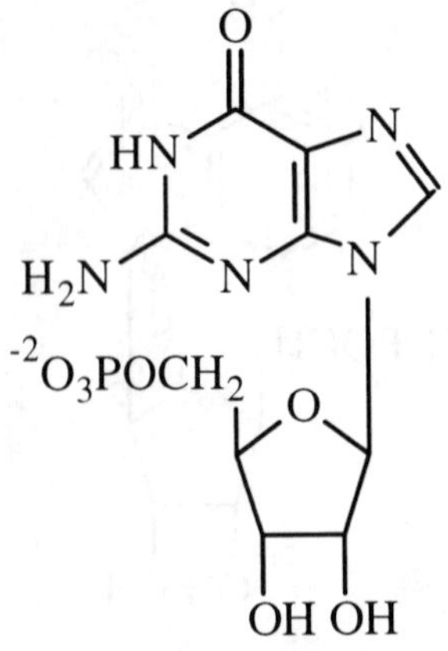

deoxyguanosine
5'-monophosphate
(dGMP)

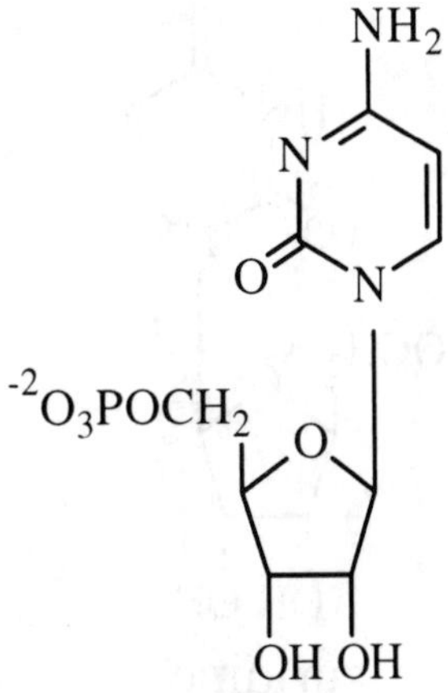

deoxycytidine
5'-monophosphate
(dCMP)

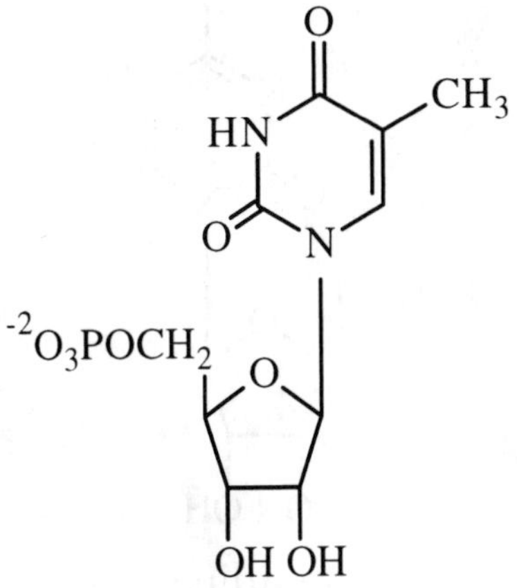

deoxythymidine
5'-monophosphate
(dTMP)

Amino acids are linked by peptide bonds to form proteins. Nucleotides are joined by phosphodiester bonds to form nucleic acids.

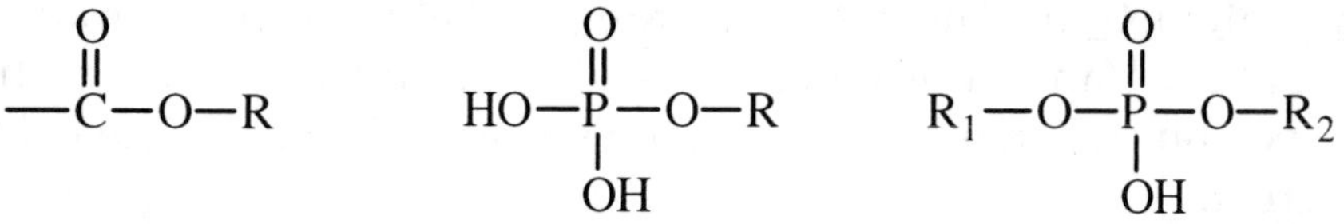

$$\underset{\text{ester}}{-\overset{\overset{\textstyle O}{\|}}{C}-O-R} \qquad \underset{\text{phosphoester}}{HO-\overset{\overset{\textstyle O}{\|}}{\underset{\underset{\textstyle OH}{|}}{P}}-O-R} \qquad \underset{\text{phosphodiester}}{R_1-O-\overset{\overset{\textstyle O}{\|}}{\underset{\underset{\textstyle OH}{|}}{P}}-O-R_2}$$

The phosphodiester bond forms between the 3'-OH of one nucleotide and the 5'-OH of the next nucleotide.

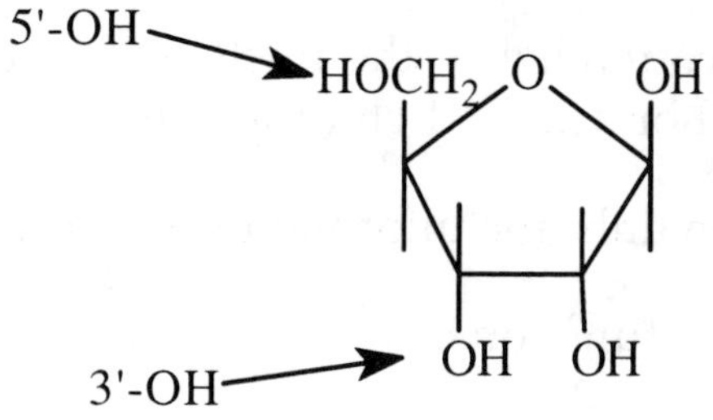

The pK_a of a phosphodiester is approximately two, so at physiological pH (7) the phosphate is negatively charged. Therefore, nucleic acids are polyanions (negatively charged). The negatively charged phosphates are usually associated with Mg^{2+} or cationic proteins (proteins with basic amino acids such as arginine or lysine).

The direction of a nucleic acid strand is given by the ends of the sugar which are not participating in a phosphodiester bond. The notation is 5' $\longrightarrow$ 3' or 3' $\longrightarrow$ 5'. Nucleic acid strands can form antiparallel double helix molecules. The base from one strand is able to hydrogen bond to a base on the opposite strand. This is referred to as a **base pair**.

Two hydrogen bonds can form between As and Ts, whereas three hydrogen bonds form between Gs and Cs. Because of the specificity of base pairing (A=T, but not with Gs or Cs), the strands are complementary and therefore serve as templates for each other. This means that if the sequence of one strand is known, then the sequence of the other strand can be determined.

Base pairs bring the two nucleic acid strands together and create a hydrophobic core. At the same time, the sugar-phosphate "backbone" twists to form a double helix.

Forces which stabilize the double helix include:

Hydrophobic: The double helix structure allows the hydrophobic purine and pyrimidine rings to reside in the interior and away from solvent.

Stacking interaction: The base pairs are parallel to each other and therefore allow Van der Waal contacts. For an individual base, the force is **very** weak. But when a DNA molecule has ≈10,000 bases, the collective force is great.

Hydrogen bonding: Specific hydrogen bonds can form between the bases.

Electrostatic interactions: The negative phosphodiester groups tend to repulse each other. This is partly why the helix is twisted. Its charge needs to be neutralized or shielded by cations (Mg^{2+}, Na^+)

Types and functions of nucleic acids include:

- Most DNA is contained in the nucleus; however, some DNA also exists in mitochondria and chloroplasts.

- DNA is used mainly for information storage.

- RNA has many functions.

Four major classes of RNA include:

1. Ribosomal (rRNA): rRNA is the most abundant class of RNA, comprising ≈80% of the total RNA of the cell. rRNAs are an integral part of ribosomes.

2. Transfer (tRNA): tRNA comprises 15% of the total RNA, and carries activated amino acids.

3. Messenger (mRNA): mRNAs represent ≈3% of the total RNA, and are copies of DNA "genes." The function of mRNA is to carry the genetic information for protein synthesis from the nucleus to the ribosomes.

4. Small RNA: This minor class of RNA is associated with RNA processing events, and possesses catalytic activity.

17 Genetic Information

Objectives

1. What is semiconservative replication?

   Semiconservative replication is DNA synthesis in which each polynucleotide strand serves as a template for the synthesis of a new strand.

2. What is a replisome and what role does it play in DNA synthesis?

   A replisome is a large complex of polypeptides which is involved in DNA replication.

3. What are the major enzymatic activities involved in DNA replication in *E. coli*?

   The prokaryotic replication process contains several basic steps, each of which requires certain enzyme activities:

   a. <u>DNA uncoiling</u>: The helicase enzymes catalyze the unwinding of duplex DNA.

   b. <u>Primer synthesis</u>: Primase catalyzes the formation of short, complementary RNA segments called primers, which are required for the initiation of DNA synthesis.

   c. <u>DNA synthesis</u>: DNA polymerases catalyze for formation of phosphodiester bonds in the synthesis of the complementary DNA strand.

   d. <u>Joining of DNA fragments</u>: Discontinuous DNA synthesis of the lagging strand requires the enzyme ligase to join the newly synthesized segments.

   e. <u>Supercoiling control</u>: The tangling of DNA strands, which can prevent further unwinding of the double helix, is prevented by the DNA topoisomerases.

4. What are Okazaki fragments?

   Okazaki fragments are the discontinuous DNA segments of the lagging strand.

5.   How does DNA replication differ in prokaryotes and eukaryotes?

The differences appear to be related to the size and complexity of eukaryotic genomes.

Timing of replication: While rapidly growing bacterial cells undergo replication throughout most of the cell cycle, eukaryotic cells are limited to a specific period of time referred to as the S phase.

Replication rate: Because of the complex structure of chromatin, DNA replication is significantly slower in eukaryotes than in prokaryotes.

Replicons: Eukaryotes have compressed the replication of their large genomes into short time periods with the use of multiple replicons.

Okazaki fragments: The fragments of eukaryotes are significantly shorter than those which occur in prokaryotes.

6.   How is DNA damage repaired?

A variety of repair mechanisms are used by the cell to repair damaged DNA. These include excision repair, photoreactivation repair, and recombinational repair.

Excision repair: Incorrect bases are removed and replaced with the correct base. Damaged DNA is detected by a distortion of the DNA. An endonuclease (called excinuclease) removes a single-stranded portion (12-13 nucleotides in prokaryotes, 27 to 29 nucleotides in eukaryotes) . The nucleotides in the excised DNA segment are replaced by DNA polymerase I and the break in the phosphodiesterase backbone sealed by DNA ligase.

Photoactivation repair: For pyrimidine dimers, an enzyme (called photoreacting enzyme) uses visible light to cleave the dimer, leaving the phosphodiesterase bonds intact.

Recombinational repair: Post-replication repair of DNA. Replication of DNA is interrupted by damaged DNA, since the replication complex dissociates from the DNA and re-initiates after the damaged site. This results in a gap in the daughter strand. This gap is repaired by an exchange of the corresponding segment of the homologous DNA (this process is called recombination). After recombination, DNA polymerase and DNA ligase complete the repair process.

7.   What are the mechanisms of excision repair and light-induced repair?

In excision repair, mutations are removed by a series of enzymes which recognize incorrect or damaged bases and replace them with the correct ones. The process begins with the detection of a distorted DNA helix by a repair endonuclease that cuts the damaged DNA and removes a single-stranded sequence of about twelve nucleotides in length. DNA polymerase I synthesizes a new sequence using the undamaged strand as a template. The process ends when DNA ligase joins the fragments.

In light-induced repair, pyrimidine dimers are restored to their original monomeric structures. In the presence of visible light, a photoreactive enzyme cleaves the dimer, leaving the phosphodiester bonds intact. Visible light provides the energy required for bond breakage.

8. What is the difference between general recombination and site-specific recombination?

General recombination requires the precise pairing of homologous DNA molecules, while site-specific recombination requires only short regions of DNA homology.

9. What are the major differences between prokaryotic and eukaryotic transcription?

Transcription in eukaryotes is significantly more complex than that of prokaryotes. In addition to requiring three RNA polymerases, the eukaryotic process requires the combined binding of numerous transcription factors before RNA polymerase can initiate transcription. Eukaryotic mRNAs are extensively processed. In contrast, prokaryotic mRNAs are usually not post-transcriptionally processed.

10. What are ribozymes?

A ribozyme is a catalytic RNA found in several organisms.

11. How do *E. coli* cells regulate lactose metabolism?

The lac operon consists of a control element and structural genes which code for the enzymes of lactose metabolism. The control element contains the promoter site, which overlaps the operator site. The promoter site is the region to which the CAP protein binds. The structural genes Z, Y, and A specify the primary structure of β-galactosidase, lactose permease, and transacetylase, respectively. β-Galactosidase catalyzes the hydrolysis of lactose, which yields the monosaccharide galactose. Lactose permease facilitates lactose transport in the cells. In the absence of inducer, the lac operon remains off because of the binding of lac repressor to the operator. When lactose becomes available, a few molecules are converted to allolactose by β-galactosidase. Allolactose then binds to the repressor, causing a change in its conformation which promotes dissociation from the operator. The lac operon remains active until the lactose supply is consumed.

Glucose is a preferred carbon and energy source for *E. coli*. In the event that the organism is exposed to both glucose and lactose, the glucose is metabolized first. Synthesis of the lac operon enzymes only occurs after the glucose is no longer available. The delay in activating the lac operon is mediated by a catabolite gene activator protein (CAP). When glucose is depleted, cAMP levels in the cell increase, and cAMP binds to CAP, allowing CAP to bind to the lac promoter. In other words, CAP exerts a positive, or activating control, on lactose.

12. By what mechanism are eukaryotic genes regulated?

Evidence currently available indicates that eukaryotic gene expression is regulated by the following mechanisms:

a. Gene rearrangements: Certain genes are regulated by gene rearrangements, in which the genomic DNA sequence is changed. An example of this type of regulation is seen in B lymphocytes with antibody production.

b. Gene amplification: During certain stages in development, the requirements for specific gene products may be so great that the genes which encode for their synthesis are selectively amplified. For example, during the early developmental stages of fertilized eggs, the demand for protein synthesis requires amplification of the rRNA genes.

c. Transcriptional control: A significant amount of gene regulation occurs through selective transcription. There appears to be two major influences on eukaryotic transcription initiation: chromatin structure and gene regulatory proteins.

d. RNA processing: Cells often use alternative RNA processing to control gene expression. For example, alternative splicing results in different forms of α–tropomyosin, a structural protein produced in various tissues.

e. RNA transport: Nuclear export signals such as capping and association with specific proteins are believed to control the transport of processed RNA molecules through nuclear pore complexes.

f. Translational control: Eukaryotic cells can respond to various stimuli (e.g., heat shock, viral infections, and cell cycle phase changes) by selectively altering protein synthesis.

General Principles

DNA Structure

DNA is a polymer of deoxyribonucleotides linked between the 3' and 5' hydroxyl groups of the deoxyribose.

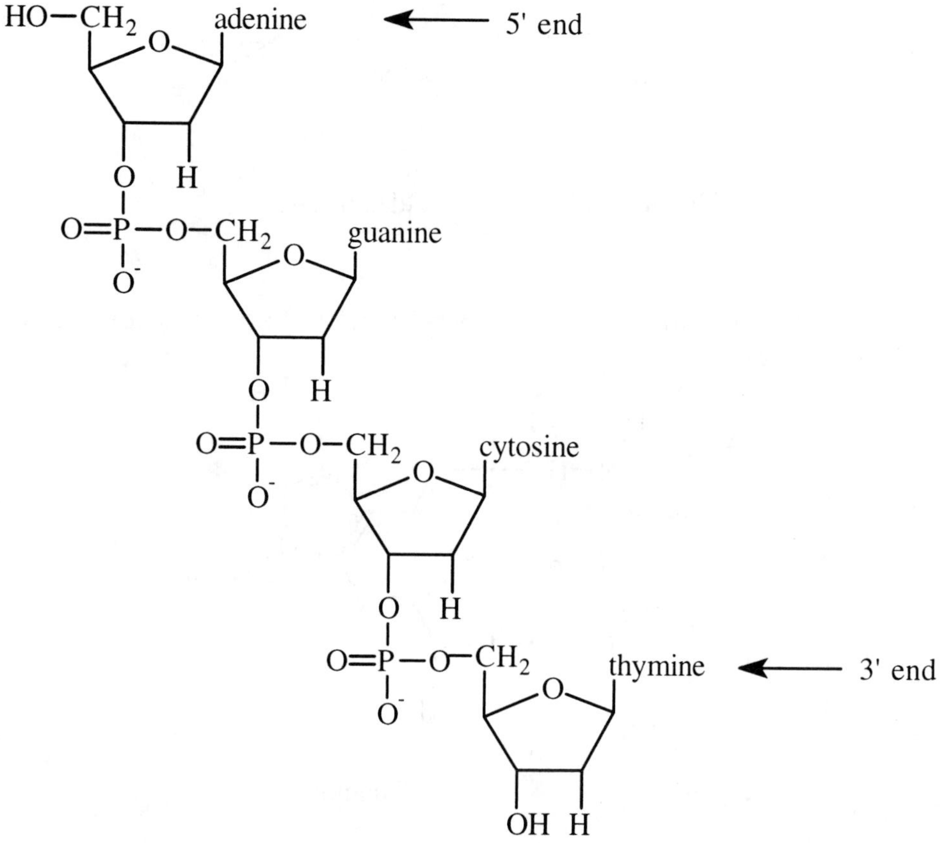

Nucleic acid sequences are written from the 5' to the 3' ends. For example, the sequence diagrammed above would be written as: dAdGdTdC or 5'-AGTC-3'.

Features of the Watson-Crick DNA Double Helix:

1. Two polynucleotide chains are coiled around each other.

2. Chains run in opposite directions or are antiparallel.

3. Nucleotide bases are on the inside of the helix, whereas the phosphate and sugar are on the outside.

4. The chains are held together by hydrogen bonds between the bases called **base pairing**.

Two hydrogen bonds form between thymine (T) and adenine (A) bases. This is noted as: T=A

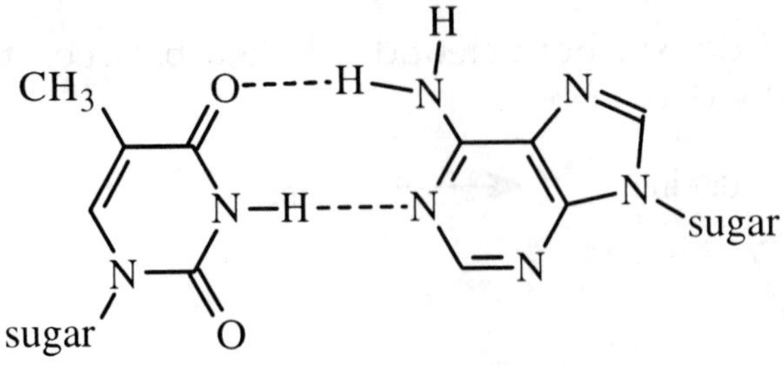

Thymine (T)   Adenine (A)

Three hydrogen bonds can form between cytosine (C) and guanine (G) bases. This is noted as: G≡C.

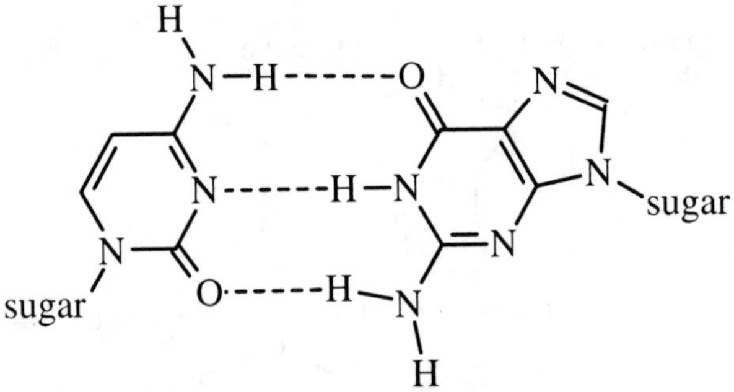

Cytosine (C)   Guanine (G)

Base pairing allows DNA to transfer information. Complementary copies can be made from a single strand. One strand can serve as the template for synthesis of the other strand.

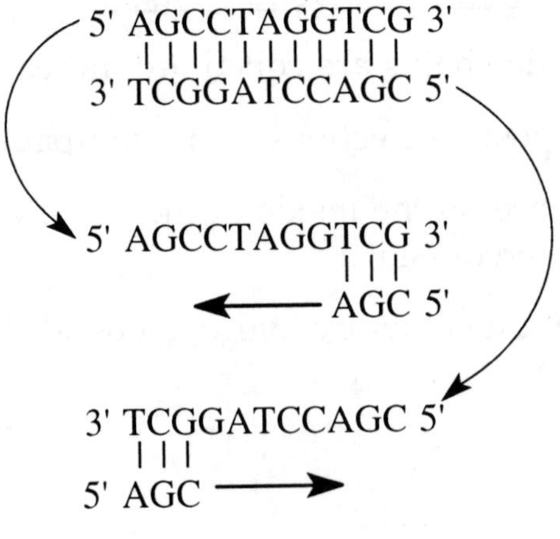

5' AGCCTAGGTCG 3'
   | | | | | | | | | | |
3' TCGGATCCAGC 5'

5' AGCCTAGGTCG 3'
   | | |
←——— AGC 5'

3' TCGGATCCAGC 5'
   | | |
5' AGC ——→

The sequence has to be complementary since As base pair only with Ts and Gs base pair only with Cs. Other combinations are not possible due to:

1. steric: A purine-purine base pair, like A=A, G=G, or A=G, would be bulky and cause the helix to be distorted. Likewise, a pyrimidine-pyrimidine base pair, such as C=C, T=T, or C=T, would also distort the double helix.

2. hydrogen bonding: Ts cannot effectively hydrogen bond with Gs, and the same holds true for Cs and As. Try this out for yourself. Draw out the bases and see if you can get them to hydrogen bond to each other.

Making DNA copies of DNA is a process called **replication** and is catalyzed by the enzyme **DNA polymerase**. DNA polymerase catalyzes the addition of deoxyribonucleotide units to a DNA chain.

$$(DNA)_n + dNTP \longrightarrow (DNA)_{n+1} + PP_i$$

Properties/Requirements of DNA polymerase:

- Deoxyribonucleotide triphosphates (dNTPs) such as dATP, dGTP, dCTP, and dTTP, **and** Mg^{2+} are the substrates for DNA polymerase.

- Deoxyribonucleotides are added to the 3'-hydroxyl of a pre-existing nucleic acid chain (either DNA or RNA).

- A DNA template is essential.

- Elongation proceeds only in the 5' ⟶ 3' direction.

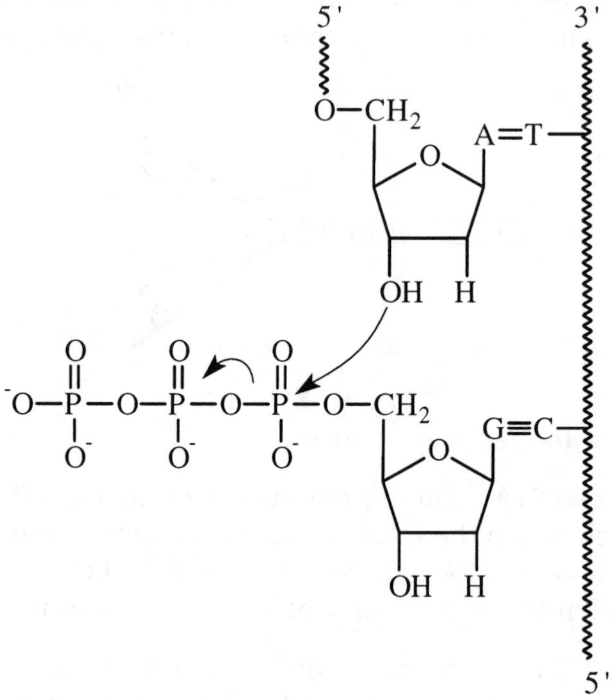

151

Many DNA polymerases have other enzymatic activities as well, such as:

- 3' ⟶ 5' Exonuclease activity. Function: "Proofreading." If the polymerase adds an incorrect base to the newly synthesized strand, the 3' ⟶ 5' exonuclease activity will remove the mistake and allow the polymerase to try again. Some DNA polymerases can also hydrolyze nucleotides from the 3' end, but only when there is a free 3'-hydroxyl group and when the nucleotide is not part of the double helix (not base paired).

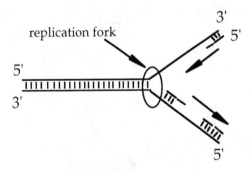

- 5' ⟶ 3' Exonuclease activity. Function: Removing primers or repairing damaged DNA. The nucleotide which is to be removed must be part of the double helix. The polymerase removes the nucleotides in front of it.

DNA replication

As a result of the 5' ⟶ 3' direction of DNA synthesis, one strand will be synthesized continuously (this is called the **leading strand**), while the other strand, the **lagging strand**, will be discontinuously synthesized as fragments.

Three steps occur at the replication fork:

1. Unwinding of DNA. Three proteins are involved. They are: **DNA gyrase**, which twists the DNA ahead of the replication fork; **Helicase**, which unwinds the DNA at the fork; and **SSB** (single-stranded binding protein), which stabilizes and protects single-stranded DNA segments.

2. Providing a 3'-hydroxyl group for DNA polymerase to add nucleotides. **Primase** is a template-dependent RNA polymerase which makes an

152

RNA primer. Primase, like most RNA polymerases, does not require a free 3'-OH group to begin synthesis.

3. In *E. coli*, DNA synthesis is provided by DNA polymerase III. This enzyme is processive, meaning it catalyzes the formation of thousands of phosphodiester bonds before dissociating from the template. The rate of synthesis is on the order of 1000 nucleotides per second.

The synthesis of the leading strand is straightforward since the synthesis of the new strand proceeds toward the replication fork. As the replication fork opens up, the leading strand is synthesized in a continuous fashion.

The synthesis of the lagging strand is a bit more complex. Because DNA synthesis can **only** proceed in the 5' $\longrightarrow$ 3' direction, a new RNA primer needs to be synthesized as the replication forks open up. Therefore, the lagging strand will be synthesized in short segments or as fragments termed Okazaki fragments. DNA polymerase III is the enzyme which catalyzes this reaction.

The RNA primers used in DNA replication need to be removed and replaced by DNA. This is where the 5' $\longrightarrow$ 3' exonuclease activity of DNA polymerase I comes into play. DNA polymerase I will degrade the RNA primers ahead of it as the enzyme synthesizes DNA. RNA primer sequences are placed with DNA sequences.

However, there is still a gap between the newly synthesized strands. This gap is closed by another enzyme called **DNA ligase**. This enzyme catalyzes the formation of a new phosphodiester bond.

RNA: Structure, function, and synthesis

The bases in RNA are:

Adenine and guanine (purines)

and

Uracil and cytosine (pyrimidine)

The sugar of RNA is ribose, so there is a 2'-hydroxyl group on the sugar.

There are three types of RNA structures

1. Transfer RNA (tRNA)

2. Ribosomal RNA (rRNA)

3. Messenger RNA (mRNA)

Each type of RNA has a different function in the cell.

Common features of all RNA types are:

- All RNAs are made from DNA templates by **RNA polymerase**

- Ribonucleotide triphosphates (ATP, GTP, UTP, CTP) are the substrates for synthesis

- Synthesis proceeds in the 5' $\longrightarrow$ 3' direction

- No primer (free 3'-hydroxyl) is required

- Only one strand of the original DNA template is copied

Transfer RNA (tRNA)

Transfer RNA represents 10-15% of the total RNA of a cell. They are single-stranded polynucleotide chains, usually about eighty nucleotides in length. tRNA functions as adapter molecules in protein synthesis. They convert a nucleotide sequence into a particular amino acid. Hydrogen bonding (or base pairing) within a tRNA molecule defines a precise three-dimensional structure. Some of the bases of tRNAs are modified and include:

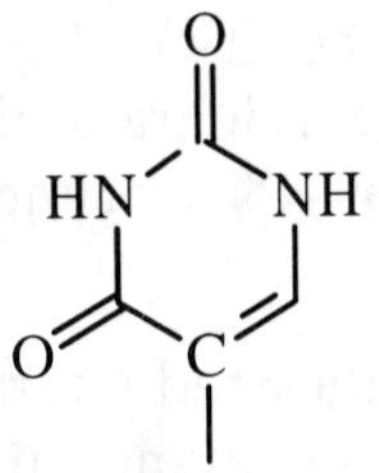

pseudouridine
(note the base is attached to
the sugar via a carbon atom)

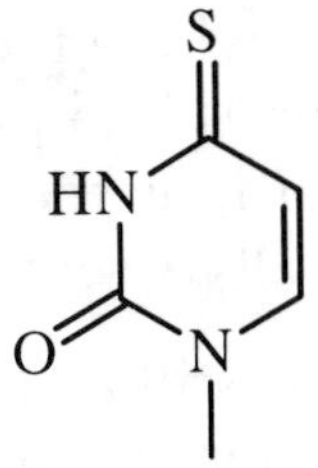

4-thiouridine
(note a sulfur atom is
present instead of oxygen)

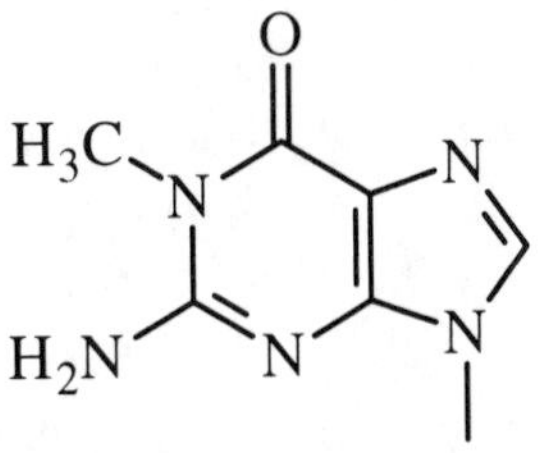

1-methyl guanosine

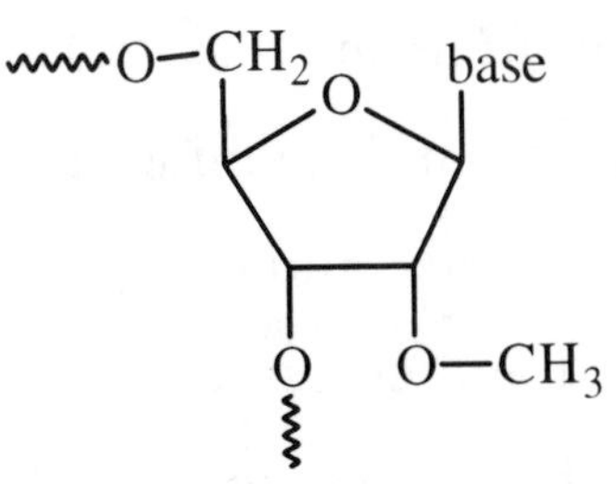

2' O-methyl ribose
(eukaryotes only)

These modifications help stabilize and protect tRNAs from degradation. These modifications can also occur in rRNAs.

Ribosomal RNA (rRNA)

Ribosomal RNAs are the most abundant type of RNA, representing 75-80% of total RNA. rRNAs are very large molecules and an integral component of ribosomes. In fact, ribosomes are 60-65% rRNA. The remainder is protein.

The structure of ribosomes differs for prokaryotes and eukaryotes. Some differences include size and composition. Prokaryotic ribosomes are slightly smaller than eukaryotic ribosomes. The difference between prokaryotic and eukaryotic ribosomes allows certain antibiotics to kill bacteria (by inhibiting protein synthesis) with little effect on the eukaryotic cell. Some of these antibiotics include:

- Chloramphenicol, which binds to the 50S ribosomal subunit.

- Gentamycin, which binds to the L6 protein of the 50S ribosomal subunit.

- Kasugamycin, which alters the methylation of the 16S rRNA thereby altering the 30S ribosomal subunit.

- Streptomycin, which binds to the S12 protein of the 30S ribosomal subunit.

Messenger RNA (mRNA)

Messenger RNA is one of the least abundant forms of RNA, representing 5-10% of total RNA. mRNA specifies the order of amino acids in a protein.

mRNA in eukaryotes is extensively processed or modified. These modifications include:

1. Splicing

2. Capping of the 5' end

3. Polyadenylation of the 3' end

These modifications, which occur in the nucleus, assist in transporting of the mRNA out of the nucleus and in helping stabilize the mRNA molecule.

Splicing of eukaryotic mRNA is necessary since not all of the DNA sequence in eukaryotes code for amino acids. Some of the genomic DNA is used to regulate DNA transcription; some serve a more structural role. **Introns** are DNA sequences which intervene and occur within a particular gene. The DNA sequence which designates the protein sequence is called the coding sequence or **exons**. Therefore to construct a contiguous nucleotide sequence to be translated into a protein sequence, removal of introns is necessary. Their removal from the newly transcribed RNA is called **splicing**.

5' capping occurs with the addition of 7-methyl guanosine to the 5' phosphate end of the eukaryotic mRNA.

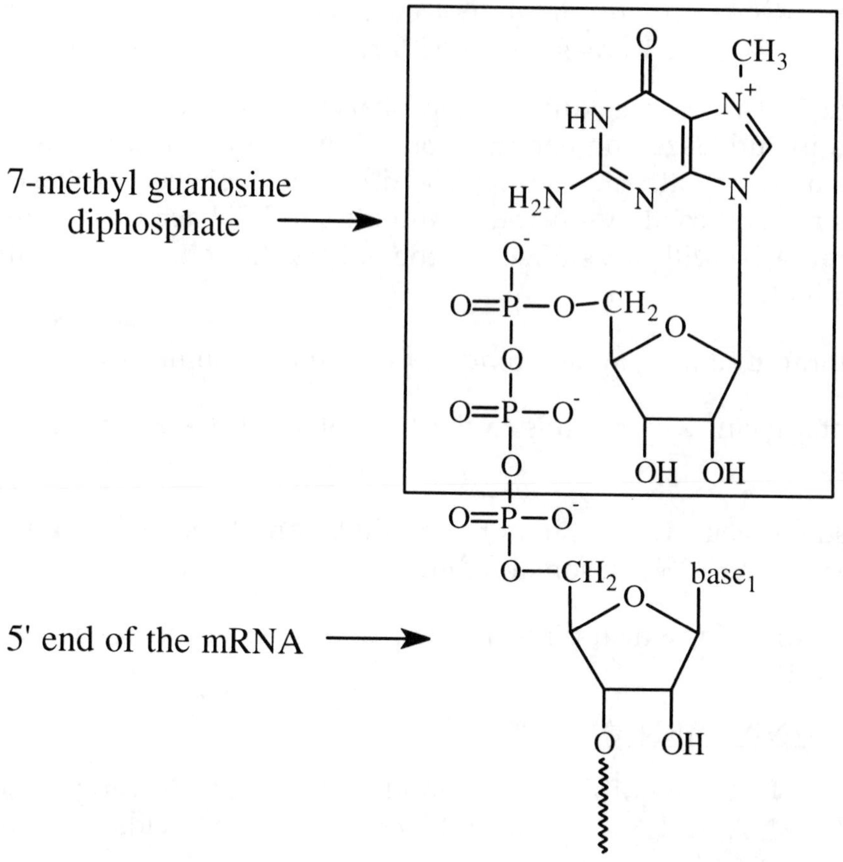

Polyadenylation of the 3' end is catalyzed by the enzyme poly(A) polymerase. This RNA polymerase does not require a template, and is therefore referred to as a template-independent RNA polymerase. Poly(A) polymerase uses ATP as the substrate and can add 100-200 AMP nucleotides to the 3' end of the mRNA.

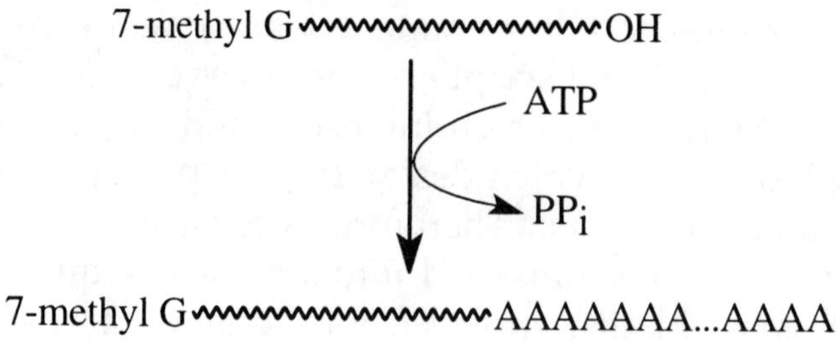

18   Protein Synthesis

Objectives

1.   What is the RNA world hypothesis?

The RNA world hypothesis is an explanation of how life first evolved. One of the principle issues of biochemical evolution is whether DNA or protein appeared first. The information encoded in DNA is required to direct the synthesis of proteins. However, certain proteins are required for DNA replication. According to the RNA world hypothesis, neither DNA nor protein came first. Instead, the first living cells which evolved contained a molecule that possessed both informational and catalytic properties, namely RNA. In this paradigm, RNA molecules were capable to some degree of self-replication and possessed an ability to catalyze primitive reactions. Later, as living cells became more complex, DNA was acquired to serve as more stable information storage molecules.

2.   What is the genetic code and what role does it play in gene expression?

The genetic code can be described as a coding dictionary which specifies a meaning for each base sequence. The genetic code thus provides a mechanism for the translation of nucleotide base sequences (gene expression) into the primary sequence of polypeptides.

3.   What is the wobble hypothesis?

The wobble hypothesis, which allows for multiple codon-anticodon interactions by individual tRNAs, is based principally on the following observations:

a. The first two base pairings in a codon-anticodon interaction confer most of the specificity required during translation.

b. The interactions between the third codon and anticodon nucleotides are less stringent.

4.   Why does the functioning of the aminoacyl-tRNA synthetases constitute a second genetic code?

The attachment of amino acids to tRNAs is a process that is catalyzed by a group of enzymes called the aminoacyl-tRNA synthetases. The aminoacyl-tRNA synthetases are a diverse group of enzymes which vary in molecular weight, primary sequence, and number of subunits. Despite this diversity, each enzyme efficiently produces a specific aminoacyl-tRNA product in a relatively error-free manner. The precision with which these enzymes esterify each amino acid to the correct tRNA is now believed to be so important for accurate translation that their functioning has been referred to collectively as the second genetic code.

5. What are the principle phases of translation?

    Despite its complexity and the variations which occur among species, the translation of a genetic message into the primary sequence of a polypeptide can be divided into three phases:

    a. Initiation: Translation begins with initiation, when the small ribosomal subunit binds to an mRNA.

    b. Elongation: It is during the elongation phase that the polypeptide is actually synthesized according to the specifications of the genetic message.

    c. Termination: During termination the polypeptide chain is released from the ribosome.

6. What role does polypeptide targeting play in cellular metabolism?

    Polypeptide targeting allows the protein to be directed to a specific location. For example, the peptide may be directed to a variety of organelles (e.g., mitochondria, chloroplasts, lysosomes, and peroxisomes).

7. What are the principle differences between prokaryotic and eukaryotic translation process?

    Some notable differences are listed below:

    Size and composition of the ribosomes: Prokaryotic ribosomes are 70S in size and composed of a 50S large subunit and a 30S small subunit. Eukaryotic ribosomes are 80S in size and composed of a 60S large subunit and a 40S small subunit. Additionally, the number and size of the rRNAs and ribosomal proteins are different.

    Initiation: Initiation of prokaryotic protein synthesis occurs by the binding of the mRNA to the 30S ribosomal subunit. This binding is facilitated by a sequence called the Shine-Dalgarno sequence, on the mRNA upstream to the start methionine. Initiation of eukaryotic translation is much more complex than for prokaryotes. Eukaryotic mRNA lacks Shine-Dalgarno sequences. Instead, eukaryotic ribosomes bind to the capped 5' end and then "scan" or migrate in the 5' ⟶ 3' direction until a translation start site is found.

    Rate of synthesis: Prokaryotic translation proceeds at a rate of 1,200 amino acids per minute, whereas eukaryotic protein synthesis is much slower at about fifty amino acids per minute.

8. What types of prominent post-translational modifications are found in prokaryotes and eukaryotes? What functions do these chemical alterations serve?

Prokaryotic polypeptides are known to undergo several types of covalent alterations:

a. Proteolytic processing: Several cleavage reactions may occur. These include the removal of the formylmethionine residue and signal peptide sequences.

b. Glycosylation: This involves the covalent linking of carbohydrate molecules to proteins, and although a few examples of glycoproteins have been found, the mechanism and functional significance are unknown.

c. Methylation: Refers to the addition of methyl groups to amino acid residues of proteins. This modification is reversible and regulates protein function.

d. Phosphorylation: Protein phosphorylation/dephosphorylation can control metabolic pathways and signal transduction in prokaryotes.

The most prominent eukaryotic post-translational changes include the alterations of prokaryotic cells and the following:

a. Hydroxylation. Hydroxylation of the amino acids proline and lysine is required for the structural integrity of the connective tissue proteins and elastin.

b. Lipophilic modifications. The covalent attachment of lipid moieties to proteins enhances membrane association and certain protein-protein interactions.

c. Disulfide bond formation. Disulfide bonds are generally found in secreted proteins and certain membrane proteins. Disulfide bridges are covalent bonds which confer considerable structural stability to the molecules which contain them.

Post-translational modifications increase the structural and functional diversity of proteins.

9. What is the significance of the signal hypothesis?

The signal hypothesis was proposed to explain the translocation of polypeptides across membrane. Its significance is that it helps explain the ability of proteins to be specifically targeted to their proper location. The fate of a targeted polypeptide depends on the location of the signal peptide and other signal sequences.

10.	How do prokaryotes and eukaryotes appear to control translation?

The extent of control of prokaryotic translation is very minor. Protein synthesis in prokaryotes is controlled predominantly by the rate of transcription.

In contrast, a wide variety of eukaryotic translational controls have been observed. In eukaryotes these mechanisms appear to occur on a continuum, from global to specific controls. Although most aspects of eukaryotic translational control are currently unresolved, the following features are believed to be important:

a.	mRNA export: The spatial separation of transcription and translation that is afforded by the nuclear membrane appears to provide eukaryotes with the opportunity to control translation.

b.	mRNA stability: In general, the translation rate of any mRNA species is related to its abundance, which in turn is dependent on both its rates of synthesis and degradation.

c.	Negative translational control: The translation of some mRNAs is known to be controlled by the binding of repressor proteins to the 5' ends of the mRNA. This effectively blocks ribosome binding and scanning.

d.	Initiation factor phosphorylation: The phosphorylation of eIF-2 in response to certain stimuli (e.g., heat shock, viral infections, and growth factor deprivation) has been observed to result in a general decrease in protein synthesis.

e.	Translational frameshifting: This process, often observed in retroviruses, allows the synthesis of more than one polypeptide from a single mRNA.

11.	What functions do molecular chaperones serve?

It has become increasingly clear that protein folding and targeting in living cells are aided by a group of molecules now referred to as molecular chaperones. These chaperone proteins bind to denatured or unfolded proteins and assist in adoption of the correct three-dimensional structure.

General Principles

Genetic code

DNA and mRNA sequences are linear polymers of four nucleotides. How does the cell convert information from four nucleotides into twenty different amino acids?

In the early 1960s, scientists concluded that groups of nucleotides must somehow designate the amino acid sequence of a protein. Using simple mathematic combinations, the genetic code was proposed to be triplets of nucleotides. For example, if the genetic code were pairs, then there would be 4 × 4 or sixteen combinations of four nucleotides (GG, GA, GT, GC, etc.). But since there are twenty amino acids, pairs of nucleotides would not provide enough combinations to encode all of the amino acids. However, using a triplet code there are 4 × 4 × 4 or sixty-four possible combinations, more than enough for twenty amino acids.

How was the genetic code elucidated?

RNA polymers were synthesized in the early 1960s with a defined sequence. For example, poly(U) or UUUUU...UUU, poly(G), poly(C), and poly(A). These RNAs were mixed with a protein synthesis extract consisting of ribosomes, charged tRNAs, etc. The resulting polypeptides were isolated and their sequence determined. From these experiments, it was determined that poly(U) gave rise to poly(phe), poly(A) ⟶ poly(lys), poly(C)⟶ poly(pro), and poly(G) ⟶ poly(gly). Repeating co-polymers such as GUGUGUGU..., AAGAAGAAG..., and GUUGUUGUU... were also synthesized, and by 1966 all possible combinations were tested and the genetic code solved.

Summary:

- All codons contain three successive, non-overlapping nucleotides.

- Many amino acids are specified by more than one codon (called degeneracy).

- Only sixty-one of the sixty-four possible codons are used to specify amino acids. The remaining three codons signal chain termination (called stop codons).

Since the genetic code is formed by successive, non-overlapping nucleotides, the actual protein sequence will depend on the starting point for the first triplet. Each potential starting point defines a unique **potential** protein sequence. This gives rise to the concept of "reading frames."

Consider the following sequence: 5' AGGCAGAACUAACCAGGUCUA 3'

frame 1:   AGG CAG AAC UAA CCA GGU CUA
            arg gln asn stop

frame 2:   A GGC AGA ACU AAC CAG GUC UA
              gly arg thr ile gln val

frame 3:   AG GCA GAA CUA ACC AGG UCU A
               ala glu leu thr arg ser

Mutations are changes in the DNA sequence which can alter the protein sequence.

Types of mutations

Substitutions

One base is replaced by another.

- transitions

    purine for purine (A ⟶ G or G ⟶ A)

    pyrimidine for pyrimidine (C ⟶ T or T ⟶ C)

- transversions

    purine for pyrimidine (C ⟶ A or C ⟶ G)

    pyrimidine for purine (A ⟶ C or G ⟶ T)

Insertions or Deletions

One or more nucleotides are inserted or deleted from the DNA sequence.

Mutations can have a variety of consequences on the protein sequence.

Non-sense mutation

This occurs when a mutation converts one codon into a stop codon. This produces premature termination of the protein.

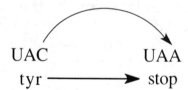

UAC   UAA

tyr ⟶ stop

Missense mutation

This occurs when a mutation converts the codon for one amino acid into a different amino acid.

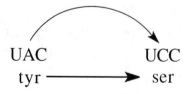

UAC   UCC

tyr ⟶ ser

Silent mutation

A process whereby a mutation does not affect the amino acid sequence due to degeneracy of the genetic code.

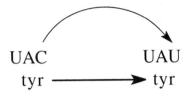

Frame shift mutation

This occurs when the insertion or deletion of nucleotide(s) change the triplet groupings. This type of mutation will change the reading frame of a protein and can have drastic effects on the protein sequence.

Protein Biosynthesis

Requires ribosomes, mRNA, tRNA, amino acids, energy, and many protein factors. The steps in prokaryotic protein biosynthesis are better understood, and unless otherwise noted, the details presented here focus on prokaryotic translation.

Four major steps:

1. Amino acid activation (charging of tRNAs or attaching an amino acid to the appropriate tRNA). Requires: amino acids, tRNAs, amino acyl tRNA synthetases, ATP, and Mg^{2+}.

2. Chain initiation (formation of the ribosome on the mRNA). Requires: fmet-tRNA, initiation codon, 30S and 50S ribosomal subunits, initiation factors (IF1, IF2, IF3), GTP, and Mg^{2+}.

3. Chain elongation (translation of the mRNA sequence into a protein sequence). Requires: 70S ribosome, amino acyl tRNAs (charged tRNAs), elongation factors (EF-Tu, EF-Ts, EF-G), GTP, and Mg^{2+}.

4. Chain termination (dissociation of the ribosome/mRNA complex). Requires: termination codon (UAA, UAG, UGA), release factors (RF1, RF2, RF3), GTP, and Mg^{2+}.

Step 1: Amino acid activation

Catalyzed by amino acyl tRNA synthetase in two parts.

The amino acid is first activated.

$$\text{amino acid} + \text{ATP} \longrightarrow \text{amino acyl-AMP} + \text{PP}_i$$

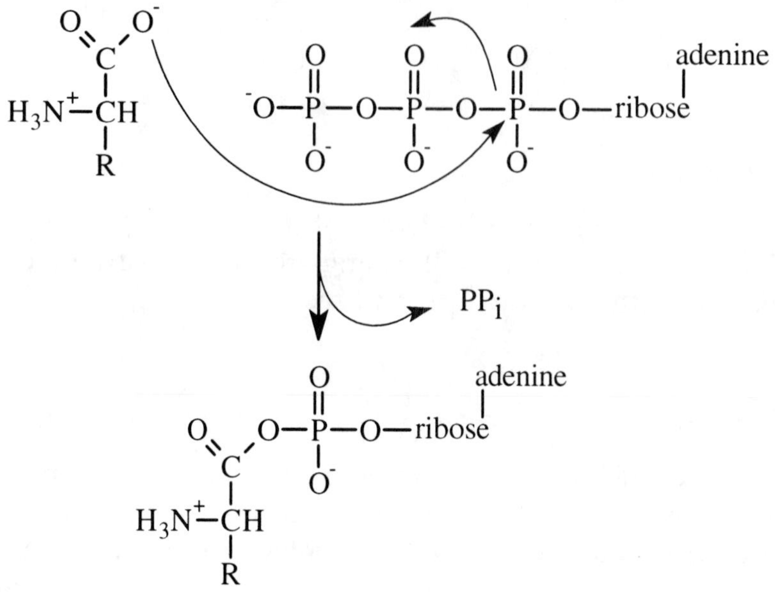

Then, the activated amino acid is transferred to the tRNA.

$$\text{amino acyl-AMP} + \text{tRNA} \longrightarrow \text{amino acyl tRNA} + \text{AMP}$$

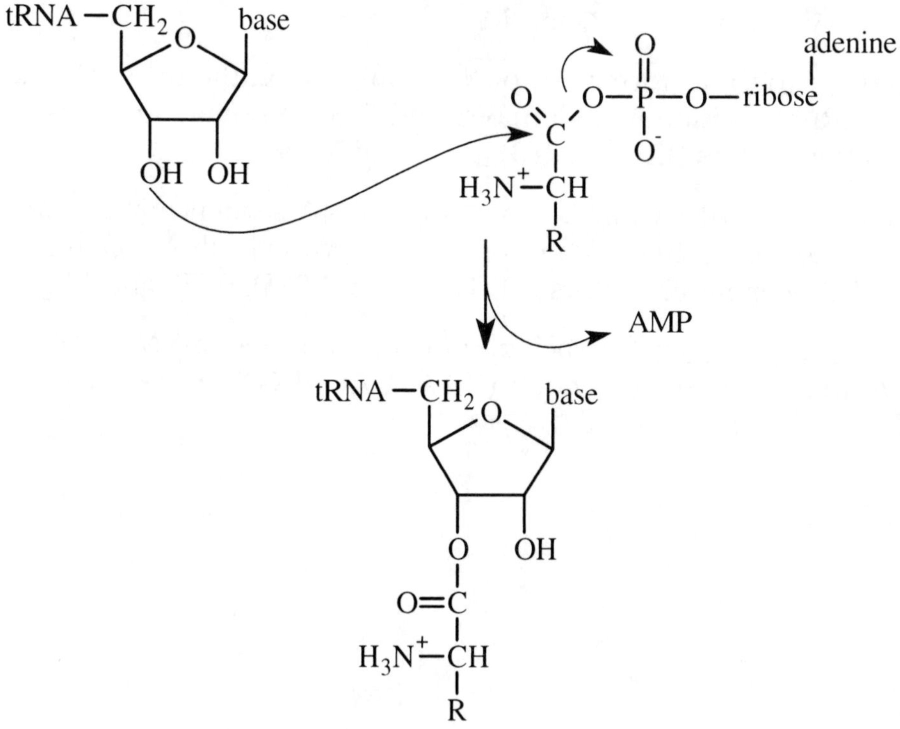

Step 2:   Chain initiation

For prokaryotes, the start of translation is indicated by a codon for methionine (AUG). Approximately ten nucleotides before the AUG codon is a nucleotide sequence (GGAGGU) known as the Shine-Dalgarno sequence. A complementary sequence to the Shine-Dalgarno sequence is found in the 16S rRNA (part of the 30S ribosome). This sequence allows the 30S ribosomal subunit to bind to the mRNA molecule.

The first amino acid in prokaryotic proteins is N-formyl methionine (fmet). There are two tRNAs for methionine:

1. fmet-tRNA, which is used to start protein synthesis.

1. met-tRNA, which is used for methionines internal to the protein sequence.

N-formyl methionine is synthesized on the charged tRNA.

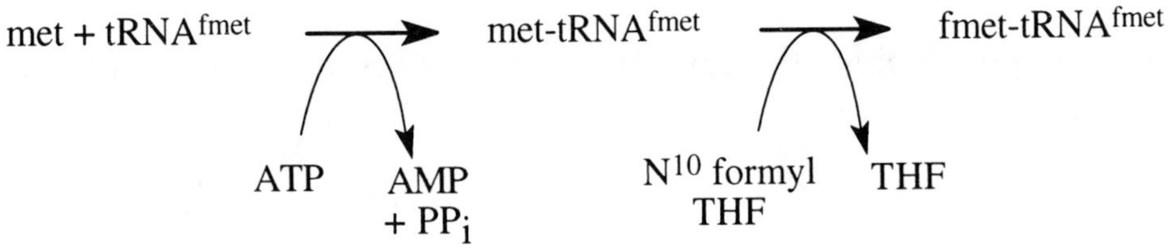

$$met + tRNA^{fmet} \longrightarrow met\text{-}tRNA^{fmet} \longrightarrow fmet\text{-}tRNA^{fmet}$$

ATP   AMP + PP$_i$

N^{10} formyl THF   THF

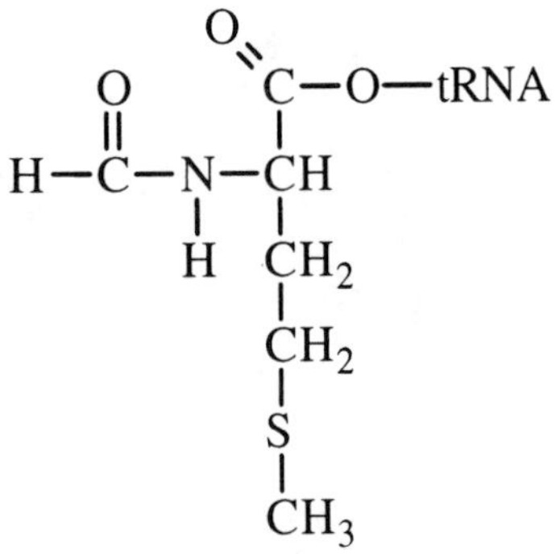

N-formyl methionine tRNA

Both tRNAs for methionine have the same anti-codon (5'-CAU-3') but tRNA^{fmet} is charged with methionine. The methionine then gets modified. tRNA^{met}, on the other hand, is charged with methionine but is not modified.

Initiation occurs in three stages:

The first stage is dissociation of the 70S ribosome into the 30S and 50S subunits. This requires two initiation factors, IF1 and IF3.

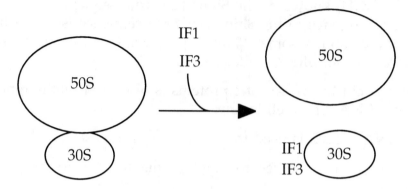

The second stage involves the formation of the 30S initiation complex.

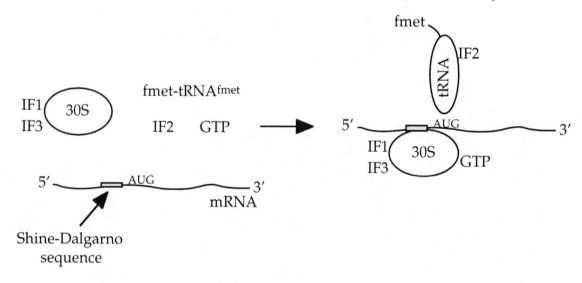

The third stage is the formation of the 70S complex.

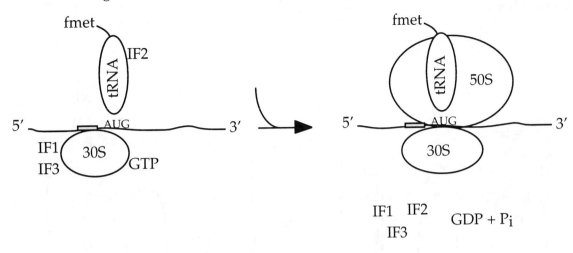

Step 3:   Chain elongation

The rate of elongation is approximately twenty to forty amino acids per second and occurs in three stages.

The first stage involves amino acyl tRNA binding. The amino acyl tRNA needs to enter the empty A-site of the ribosome complex. This process requires several accessory proteins. The amino acyl tRNA forms a complex with EF-Tu bound with GTP. This complex allows binding of the amino acyl tRNA to the ribosome's A-site. Once the EF-Tu (GTP):aa-tRNA complex binds to the ribosome, GTP is hydrolyzed to GDP + P_i. The next several reactions recycle EF-Tu bound with GDP, denoted as EF-Tu (GDP), to EF-Tu (GTP).

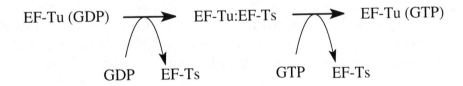

The second stage of elongation is transpeptidation or the formation of the peptide bond.

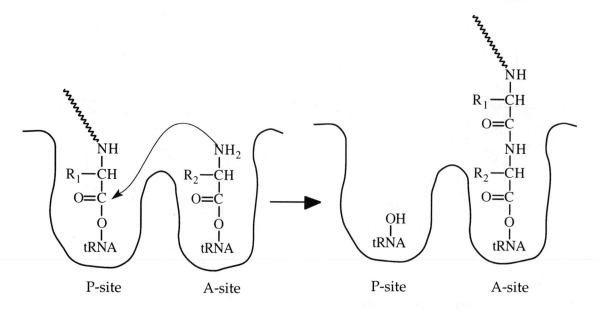

The third stage of elongation is translocation. In this process, the uncharged tRNA in the P-site must be expelled. The peptide-tRNA is moved from the A-site to the P-site, and the ribosome moves so that the next codon on the mRNA is in the A-site. This stage requires EF-G and GTP hydrolysis.

The elongation process repeats until the ribosome reaches a stop codon (UAA, UGA, UAG). The final step of protein synthesis then occurs.

Step 4:   Chain Termination

There are no tRNAs for stop codons. Instead, proteins called release factors (RFs) are used. RF1 recognizes the UAA and UAG codons, and RF2 recognizes UAA and UGA. A third RF, RF3, binds GTP and facilitates RF1 and RF2 binding to the A-site. The GTP bound to RF3 is hydrolyzed, thereby releasing the newly formed peptide 70S ribosome and the mRNA.

Features of eukaryotic translation:

Initiation: Many more (> 10) IFs, designated eIFs. There is also no Shine-Dalgarno sequence.

Elongation: Very similar to prokaryotes. eEF1 is a two subunit protein which contains both EF-Tu and EF-Ts activities. eEF2 is analogous to EF-G.

Termination: Only a single release factor, eRF, is involved.

Because of the differences in prokaryotic and eukaryotic protein synthesis, many drugs are selective inhibitors.

Many antibiotics inhibit prokaryotic protein synthesis. For example, streptomycin binds to the S12 protein of the 30S ribosomal subunit, and tetracycline inhibits protein synthesis by preventing binding of aminoacyl tRNAs to the A-site of the ribosome.

Other drugs and toxins can inhibit only eukaryotic protein synthesis. Cycloheximide blocks 60S ribosomal translocation, while diphtheria toxin inhibits eEF2. In fact, one of the most potent toxins, ricin toxin, removes a single adenine residue from the rRNA of the eukaryotic 60S ribosomal subunit. This modification completely inactivates the ribosome so that a single toxin molecule can kill a cell.

19   Biotransformation

Objectives

1.   What types of molecules are considered to be primary or secondary metabolites?

Primary metabolites are molecules such as sugars, amino acids, fatty acids, and the intermediates of the energy-generating pathways. These molecules are absolutely required for maintenance of the living state. Secondary metabolites are derived from primary metabolites. Examples of secondary metabolites are found among the alkaloids, the terpenes, various glycosides, and numerous fatty acid and amino acid derivatives.

2.   What is biotransformation and what purpose does it serve?

Biotransformation is a series of enzyme-catalyzed processes in which toxic substances are converted into less toxic metabolites. In mammals, biotransformation is used principally to convert toxic molecules, which are usually hydrophobic, into water-soluble derivatives so that they may be easily excreted. In plants, biotransformed derivatives are usually safely stored in the vacuole or cell wall. The enzymes which catalyze the biotransformation of xenobiotics are similar to several of the enzymes involved in the disposal of hydrophobic endogenous molecules.

3.   What types of reactions occur during phase I of biotransformation?

During phase I various reactions involving oxidoreductases and hydrolyses convert hydrophobic substances into more polar molecules. The conversion of hydrophobic molecules to more polar forms is done by introducing or masking a functional group (e.g., -OH, -NH$_2$, or -SH).

4.   What types of reactions characterize phase II of biotransformation?

Phase II consists of reactions with substances such as glucuronate, glutamate, sulfate, or glutathione. Conjugation results in a dramatic improvement in solubility, which then promotes rapid excretion. During this process, lipophilic metabolites, bearing functional groups which can act as acceptors, undergo enzyme-catalyzed reactions along with second (or donor) substrates.

5. How do mammals and plants differ in their phase III processing of xenobiotics?

The excretion of biotransformed molecules in animals is sometimes referred to as phase III. Phase III in plants does not exist because the biotransformed derivatives are safely stored in the vacuole or cell wall.

6. What are the differences between the reactions catalyzed by cytochrome P450 and flavin-containing monooxygenases?

All the oxidative reactions which are catalyzed by cytochrome P450 may be viewed as hydroxylation reactions. The flavin-containing monooxygenases catalyze the oxidation of substances containing functional groups such as nitrogen, sulfur, or phosphorus.

7. What are the consequences of drug tolerance and cross-tolerance?

Tolerance is defined as a decrease in the patient's responsiveness to a drug. In other words, after taking a drug for a period of time, stronger doses would be required to achieve the equivalent degree of symptomatic relief. Cross tolerance can occur when two different drugs are detoxified by the same enzyme. As a result, both drugs will require dosage increases to maintain their clinical effectiveness. When one of the drugs is removed, the induced enzyme quickly reverts back to its previous level. This can cause the patient to overdose on that drug.

8. How is ethanol detoxified? Why is this drug considered such a serious threat to health?

The major mechanism by which ethanol is detoxified involves the cytoplasmic enzyme of alcohol dehydrogenase (ADH). ADH catalyzes the oxidation of ethanol to acetaldehyde. Acetaldehyde is converted to acetate by aldehyde dehydrogenase located within the mitochondrial matrix. Most of the acetate produced by this reaction is metabolized further in other tissues, such as cardiac and skeletal muscles.

When the concentration of ethanol in the liver becomes sufficiently high, it is also detoxified by the microsomal ethanol-oxidizing system (MEOS). The MEOS consists of an ethanol-inducible cytochrome P450 isozyme and NADPH cytochrome P450 reductase. In addition, the ethanol-inducible cytochrome P450 apparently possesses a unique capacity to convert xenobiotics (e.g., CCl_4) to toxic metabolites.

Because of the liver's important role in metabolism, liver damage caused by alcohol has widespread effects. A few of the more obvious alterations in vital hepatocyte functions caused by ethanol-induced toxicity include the following:

a. Hepatic steatosis (or fatty liver): Hepatic steatosis, the abnormal accumulation of fat within the liver, is one of the earliest manifestations of alcohol abuse. As fat accumulates, hepatocytes are increasingly unable to perform their metabolic functions.

171

b. <u>Hyperlactacidemia (high blood lactate concentration)</u>: Hyperlactacidemia results from decreased conversion of pyruvate to acetyl-CoA, a consequence of depressed citric acid cycle activity. High levels of lactate contribute to a reduction in the kidney's capacity to excrete uric acid, thereby promoting hyperuricemia.

c. <u>Malnutrition</u>: The relationship between excess alcohol consumption and malnutrition is complex. Because of ethanol's high caloric value, alcoholic beverages often displace other foods. Chronic and excessive alcohol consumption damages the gastrointestinal tract to the point where the digestion and absorption of food is compromised.

9.   What roles do plants play in the processing of environmental toxins?

Plants serve as a "sink" for environmental pollutants. Plants possess an ability to metabolize xenobiotics which bear a striking resemblance to animal biotransformation processes. Plants are able to store these biotransformed molecules in their cell walls or vacuoles.

Additional Concepts

Monoamine oxidase (MAO) is an enzyme which oxidatively deaminates many amines and drugs. The reaction products are the aryl or alkyl aldehydes, which are further metabolized by other enzymes to the corresponding carboxylic acids.

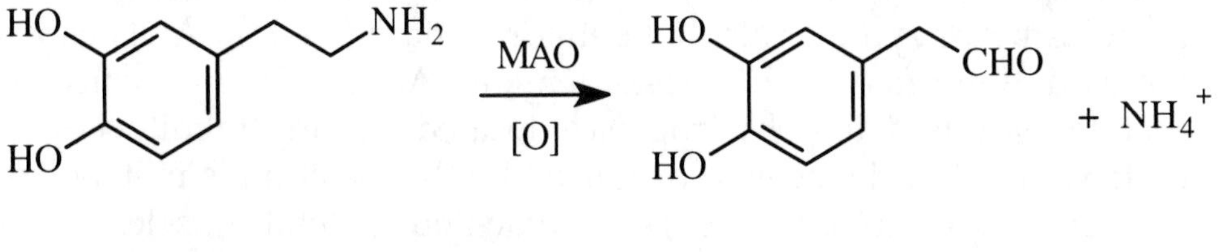

Dopamine

MAO is a mitochondrial enzyme found in the liver, kidney, intestine, and brain. Its substrates include phenylethylamines, tyramine, catecholamines (dopamine, norepinephrine, epinephrine), and tryptophan derivatives (serotonin, tryptamine). The function of this enzyme in the central nervous system (CNS) appears to be to assist in the termination of neurotransmission.

Antibacterial research to fight tuberculosis lead to the development of the hydrazide, iproniazid:

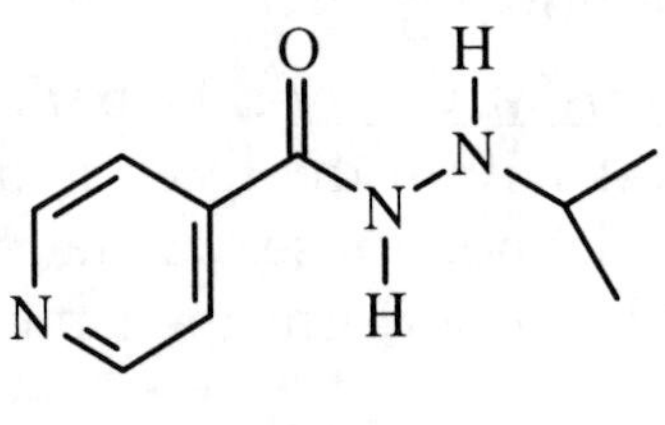

When iproniazid was clinically tested, an unexpected side effect was observed: mood elevation (euphoria). It was later found that iproniazid and other hydrazides were potent inhibitors of MAO. Other non-hydrazide inhibitors of MAO have been developed.

Tranylcypromine   Pargyline

Inhibition of MAO results in the elevation of monoamines (dopamine, serotonin) in the CNS, but how this is related to mood-elevation is uncertain. A variety of toxic effects are also associated with MAO inhibitors, such as hypotension, liver damage, nausea, vomiting, constipation, dry mouth, and delusion. Certain foods, such as cheese (e.g., Brie, Stilton, and New York cheddar) are rich in tyramine. Ordinarily, tyramine is rapidly oxidized by MAO; however, cheese toxicity was observed in patients who received the MAO inhibitor tranylcypromine.

An unfortunate case of biotransformation leads to new insight into the mechanism for Parkinson's disease.

Parkinson's disease, sometimes referred to as "shaking palsy," affects approximately a half million Americans and is the third most common neurological disease. Symptoms begin with hand tremors and ultimately lead to rigidity and the inability to initiate voluntary movement. Parkinson's disease appears to result from the death of nigral neurons in the basal ganglia. Although the mechanism for neuronal death is still unknown, insight into a possible mechanism came in 1982 when seven, young drug users in San Francisco developed the same symptoms as Parkinson's disease after they had self-administered a synthetic heroin derivative. This was highly unusual since the mean age of Parkinson's disease onset is 58 years. It was discovered that their synthetic heroin was contaminated with 1-methyl-4-phenyl-1,2,3-tetrahydropyridine (MPTP). MPTP's toxic effects required that it be biotransformed into MPP$^+$ (1-methyl-4-phenylpyridinium) by monoamine oxidase.

MPTP   MPP$^+$

MPP$^+$ is then taken into nigral neurons by the dopamine transporter and, once inside the neuron, enters the mitochondrion where it inhibits oxidative phosphorylation, resulting in neuronal death. The possibility that MAO may be involved in Parkinson's disease has led to the use of MAO inhibitors to slow the progression of Parkinson's disease as well as elevated dopamine levels.

Chapter 1 Answers to Even Numbered Questions

2. The major differences between prokaryotes and eukaryotes are: size (prokaryotes are significantly smaller), complexity (eukaryotes have a complex internal structure), and biochemical diversity (many prokaryotes possess biochemical mechanisms that allow them to exploit extremely harsh environments).

4. The major classes of small biomolecules are the amino acids, sugars, fatty acids, and nucleotides. These molecules are found in proteins, polysaccharides, fats and phospholipids, and nucleic acids, respectively.

6. Fatty acids are components of the triacylglycerols (energy storage molecules) and the phosphoglycerides (important membrane components). Sugars are used as energy sources and structural components. Nucleotides compose nucleic acids, and participate in energy-generating reactions.

8. The principal mechanism by which cells obtain energy is by the oxidation of certain nutrient biomolecules.

10. The five kingdoms of living organisms are Monera, Protista, Fungi, Plantae and Animalia. Examples of monera include bacteria such as *Escherichia coli* (an important inhabitant of the human intestinal tract), *Lactobacillus acidophilus* (used in the production of yogurt) and *Clostridium botulinum* (the causative agent of botulism), and cyanobacteria such as *Anabaena azo llae*. The fungi include various types of mushrooms, yeasts and molds. A few plant examples are oak and cypress trees, grasses such as wheat and rye, and daffodils. Examples of animals include dogs, elephants, tigers and squirrels. All living organisms are eukaryotic except the Monerans. Animals and plants are multicellular organisms.

12. The assumption that the biochemical processes in prokaryotes and eukaryotes are similar is only safe when basic living processes are considered (e.g., glycolysis and the general principles of genetic inheritance). Living organisms are so diverse in their adaptations that information acquired from research with prokaryotes must be judiciously interpreted in reference to eukaryotes.

14. (a) amino acid, (b) sugar, (c) fatty acid, (d) lipid (steroid), (e) nucleotide.

16. Organelles are subcellular structures that perform specific tasks in living cells. Organelles allow eukaryotic cells to perform sophisticated functions.

18. The primary functions of metabolism are acquisition and utilization of energy, synthesis of biomolecules, and removal of waste products.

20. Highly ordered living organisms evade disorder only by the constant acquisition and utilization of energy and building material, and the removal of waste products.

22. The common types of chemical reactions in living cells include: nucleophilic substitution, elimination, isomerization, and oxidation-reduction reactions.

24. The functions of polypeptides include transport, structural composition, and catalysis (enzymes).

26. Nucleic acids are the largest biomolecules. DNA is the repository of an organism's genetic information. RNA is involved in the expression of genetic information (primarily in various aspects of protein synthesis).

28. Order is maintained in living organisms primarily by the synthesis of biomolecules, the transport of ions and molecules across cell membranes, the production of force and movement, and the removal of metabolic waste products.

Chapter 2 Answers to Even Numbered Questions

2. All living cells have similar chemical compositions (i.e., they are all composed of molecules such as carbohydrates, proteins, and lipids) and they all utilize DNA as genetic material.

4. The plasma membranes of both prokaryotic and eukaryotic cells control the flow of substances into and out of the cell. In addition, plasma membrane receptors bind to specific molecules in the cell's external environment. In prokaryotes, for example, some receptors allow the organism to respond to the presence of food molecules. In eukaryotes, numerous cell receptors bind specific hormone or growth factor molecules. The prokaryotic cell wall is sufficiently rigid that it maintains the organism's shape and protects against mechanical injury.

6.   a.   Exocytosis is a cellular process which consists of the fusion of membrane-bound secretary organelles with the plasma membrane. The contents of the granules are then released into the extracellular space.

    b.   Biotransformation is a biochemical process in which water-insoluble organic molecules are prepared for excretion.

    c.   Grana consist of tightly stacked portions of thylakoid membrane within chloroplasts.

    d.   Symbiosis is defined as the living together of two dissimilar organisms in an intimate relationship.

    e.   Endosymbiosis is a mechanism that has been proposed to explain the evolution of modern eukaryotic cells. In the hypothesis, primordial eukaryotic cells engulfed smaller prokaryotic cells that eventually became modern mitochondria, chloroplasts and possibly other cell structures such as flagella and cilia.

    f.   Proplastids are small nearly colorless plant cell structures that develop into the plastids of differentiated cells.

    g.   The thylakoid is an intricately folded membrane system that is responsible for several chloroplast metabolic functions.

8.   Plant cells; leucoplasts; chromoplast.

10.   The highly developed framework of the cytoskeleton performs the following functions in eukaryotic cells: (1) maintenance of overall cell shape, (2) facilitation of coherent cellular movement, and (3) provision of a supporting structure that guides the movement of organelles within the cell.

12.   Without encumbering and rigid cell walls, the differentiated cells of eukaryotes can assume an astonishing variety of shapes. In addition, some eukaryotic cells can engage in ameboid-type movement.

14.   Specialized cells can perform very sophisticated functions that make multicellular organisms possible. Cell specialization can be considered a disadvantage because such cells cannot exist independently; that is, they can only exist as part of a multicellular organism where their metabolic needs (e.g., energy requirements and waste product removal) are met.

16.   The presence of DNA in the organelle would suggest that it had been at one time a free living organism.

18. The principal function of the rough endoplasmic reticulum is the synthesis of membrane proteins and protein for export from the cell. Smooth endoplasmic reticulum, so-named because it lacks attached ribosomes, is 'involved in lipid synthesis and biotransformation processes.

20. The ribosomes of mitochondria and chloroplasts are similar to those of prokaryotes. Several antibiotics that inhibit the protein synthesis activities of prokaryotic ribosomes also inhibit mitochondrial and chloroplast ribosomal function.

Chapter 3 Answers to Even Numbered Questions

2. $pH = -\log [H^+]$
$8.3 = -\log [H^+]$
$[H^+] = 10^{-8.3} = 5.0 \times 10^{-9}$ M

4. In order to prepare a 0.1 M phosphate buffer with a pH of 7.2, the Henderson-Hasselbach equation should be used to calculate the ratios of the salt and acid:

$$pH = pK_a + \log [Salt]/[Acid]$$

From a table of ionization constants the phosphate conjugate acid base pair with a pK_a closest to 7.2 is chosen:

$$H_2PO_4^- \rightleftharpoons H^+ + HPO_4^{-2}$$
$$pK_a = 7.2$$

Substituting these values into the equation gives:

$$7.2 = 7.2 + \log [Salt]/[Acid]$$
$$0 = \log [Salt]/[Acid]$$
$$10^0 = [Salt]/[Acid] \text{ or } [Salt] = [Acid] \text{ (equation 1)}$$

The concentrations of the conjugate base and acid must be equal. We also know that the total concentration of the phosphate buffer is 0.1 M. Therefore,

$$[Salt] + [Acid] = 0.1 \text{ M (equation 2)}$$

Using simultaneous equations (i.e., substituting value from equation 1 into equation 2) to determine the concentrations of the salt and acid for this particular buffer solution gives,

$$[Salt] + [Salt] = 0.1 \text{ M}$$
$$2 [Salt] = 0.1 \text{ M or } [Salt] = 0.05 \text{ M}$$

Taking this value and inserting it into equation 1, gives

$$[Acid] = 0.05 \text{ M}$$

To prepare this buffer place 0.05 mol each of the salt and acid in a 1 L volumetric flask and dilute with water to the 1 L mark.

6. Osmolarity = Molarity x the number of ions produced.
    Na$_3$PO$_4$ dissociates into four ions. Assuming 85% ionization, the osmolarity of a 1.3 molar solution of Na$_3$PO$_4$ would therefore be 4 x 1.3 x 0.85 = 4.4

8. a. 1 M sodium lactate, osmolarity = 2. Water would flow into the sucrose-containing dialysis bag.
    b. 3 M sodium lactate, osmolarity = 6. Water would flow out of the sucrose-containing dialysis bag.
    c. 4.5 M sodium lactate, osmolarity = 9. Water would flow out of the sucrose-containing dialysis bag.

10. The equation for osmotic pressure is $\pi = MRT$. Substituting values gives:

    0.01 atm = M (0.082 L atm/K mol) (298 K)

    4.1×10^{-4} mol/L = M

    The concentration is 56 mg/30 mL = 1.87 mg/mL or 1.87 g/L
    Therefore,

    $$\frac{1.87 \text{ g/L}}{4.1 \times 10^{-4} \text{ mol/L}} = 4,561 \frac{\text{g}}{\text{mol}}$$

    The molecular weight is 4,600 g/mol.

12.

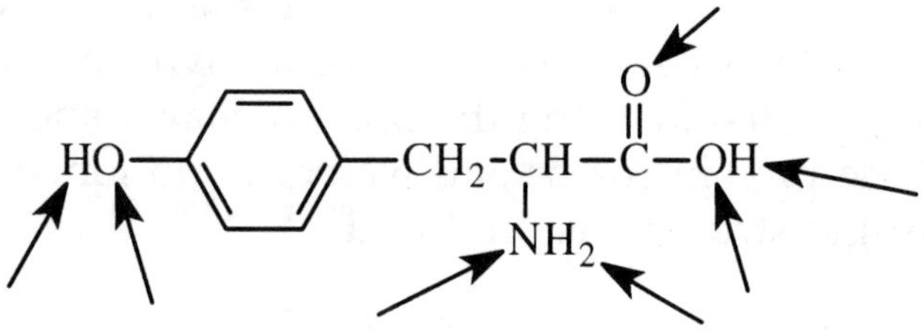

    Arrows indicate atoms that would be involved in hydrogen bonding.

14. b, c and d all would be expected to have a dipole moment.

16. Only d would form a micelle.

18. Only c would be capable of forming a buffer system.

20. Hyperventilation drives the transfer of carbon dioxide from the blood. This process, which shifts the following equilibrium, consumes protons thereby making the blood more alkaline.

$$CO_2 + H_2O \rightleftharpoons H_2CO_3 \rightleftharpoons HCO_3^- + H^+$$

$$\longleftarrow$$

22. Osmolarity is molarity multiplied by the number of ions produced as the compound dissolves.

24. The high salt content of the seawater would cause excessive amount of water to move out of the plant cells thereby killing them.

26. As steam condenses, its high heat of vaporization results in the release of a great deal of energy and a rapid and lethal increase in the temperature of any exposed organism.

28. $\Delta T_b = K_b m$
   For 2 molar Na_2SO_4 the equation becomes:

   $\Delta T_b = (0.54) (2) (3) = 3.24°C$
   The boiling point of the water would be 103.24°C

30. In a liquid the molecules must be free to move over one another. In the Jell-O solution each water molecule hydrogen bonds to two chains of the protein, locking the protein chains and the water together. The protein chains and water are no longer free to move and a semirigid structure is produced.

32. The amount of energy required to change the concentration of a solution (i.e., to remove pure water) is independent of the way the change is achieved. Therefore, both processes should use exactly the same amount of energy. However, in distillation there is always heat loss to the environment. Additional energy must be supplied to make up for this heat loss. In reverse osmosis there is no analogous pressure loss and less energy is required to purify the same amount of water.

34. The function of the cell membrane is possible only because of the insolubility of lipids in water. If water dissolved everything in living organisms, the membranes could not form and living organisms would not be possible.

Chapter 4 Answers to Even Numbered Questions

2. State functions are those that are independent of path. Entropy, enthalpy, and free energy are all path independent.

4. Given that the ionization constant for formic acid is 1.8×10^{-4}, the $\Delta G°$ for the reaction would be calculated as follows:

$$\Delta G° = -2.303 \, RT \log K_{eq}$$
$$\Delta G° = (-2.303)(1.987 \, cal/mol \, K)(298 \, K) \log (1.8 \times 10^{-4})$$
$$\Delta G° = (-2.303)(1.987)(298)(-3.75)$$
$$\Delta G° = 5114 \, cal/mol \text{ or } 5.1 \, kcal/mol$$

6.

$$CH_3CH_2OH \longrightarrow CH_3CHO + 2H^+ + 2e^- \qquad +0.2 \, V$$
$$NAD^+ + 2H^+ + 2e^- \longrightarrow NADH + H^+ \qquad -0.32 \, V$$
$$\overline{}$$
$$CH_3CH_2OH + NAD^+ \longrightarrow NADH + H^+ \qquad -0.12 \, V$$

$$\Delta G°' = -n \, F \, \Delta E_o'$$

$$\Delta G°' = -(2)(23{,}062 \, cal/V \, mol)(-0.12 \, V)$$

$$\Delta G°' = 5.5 \, kcal/mol = 6 \, kcal/mol$$

Under these conditions this reaction is endergonic.

8. Under standard conditions the following statements are true: a, c, d, e, f and g.

10. The equilibrium constant for the conversion of glucose-1-phosphate to glucose-6-phosphate is calculated as follows:

$$\Delta G° = -1.7 \, kcal/mol$$
$$\Delta G° = -2.303RT \log K_{eq}$$
$$-1700 \, cal/mol = -(2.303)(1.987 \, cal/mol \, K)(298 \, K) \log K_{eq}$$
$$1700 = 1364 \log K_{eq}$$
$$\log K_{eq} = 1700/1364 = 1.25$$
$$K_{eq} = 10^{1.25} = 17.8$$

12. AMP hydrolysis involves cleavage of an ester bond and therefore releases the least energy. Hydrolysis of the other phosphate linkages involves either the hydrolysis of an anhydride or enol bond.

14. The voltage for the reactions is calculated as follows:

$$Fe^{+2} \longrightarrow Fe^{+3} + 1e^- \qquad -0.77 \text{ V}$$

$$\underline{Pyruvate + 2e^- + 2H^+ \longrightarrow Lactate \qquad -0.19 \text{ V}}$$

$$Pyruvate + 2H^+ + 2Fe^{+2} \longrightarrow Lactate + 2Fe^{+3} \qquad -0.96\text{V}$$

16. For the reactions,

$$H_2S + 2\,O_2 \longrightarrow H_2SO_4 \qquad \Delta H_F = -189.1 \text{ kcal/mol}$$

$$6\,CO_2 + 6\,H_2O \longrightarrow C_6H_{12}O_6 + 6\,O_2 \qquad \Delta H_F = +668.8 \text{ kcal/mol}$$

All of the oxygen produced by the fixation of carbon dioxide in reaction II would make possible the production of 3 moles of sulfuric acid and -567.3 kcal/mol (3 x -189.1) of energy. This is less energy than is required to fix the carbon dioxide in reaction II (+668.8 kcal/mol). Oxygen must therefore diffuse from the ocean's surface to make possible the oxidation of hydrogen sulfide and the generation of additional energy.

18. In the case of endothermic solutions, the enthalpy may be negative but the entropy is sufficiently positive to make the overall $\Delta G°$ favorable.

20. ATP has an intermediate phosphate group transfer potential. This makes it possible for ATP to serve as a carrier of phosphoryl groups from high energy compounds to those of lower energy.

Chapter 5 Answers to Even Numbered Questions

2. a. Nonpolar, b. Polar, c. Acidic, d. Basic, e. Nonpolar, f Basic, g. Nonpolar, h. Polar, i. Nonpolar, j. Nonpolar.

4. The species present at the first plateau are:

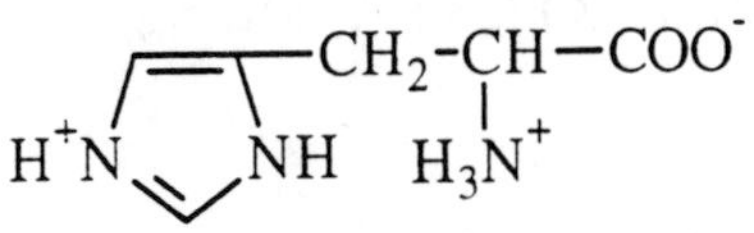

and

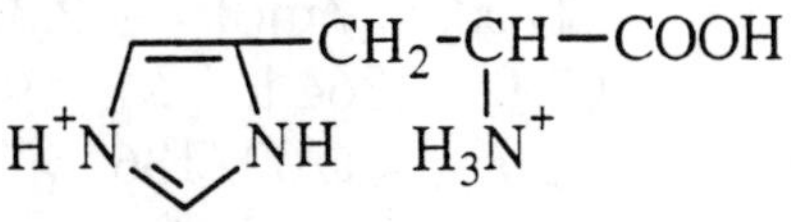

The species present at the second plateau are:

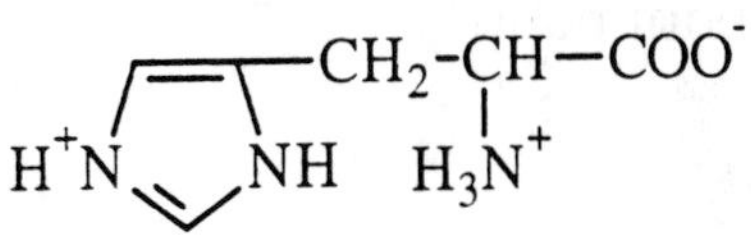

and

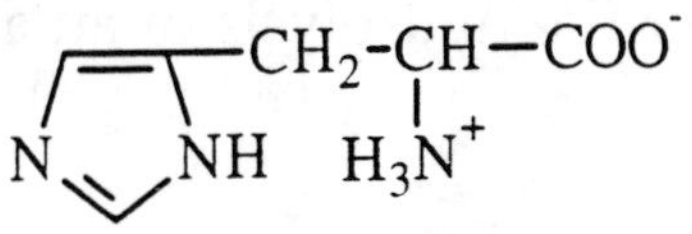

The species present at the third plateau are:

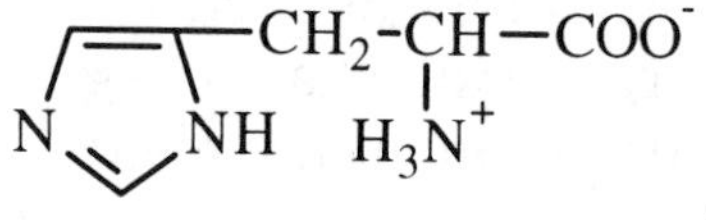

$$\text{imidazole ring}-CH_2-\underset{\underset{H_3N^+}{|}}{CH}-COO^-$$

and

$$\text{imidazole ring}-CH_2-\underset{\underset{H_2N}{|}}{CH}-COO^-$$

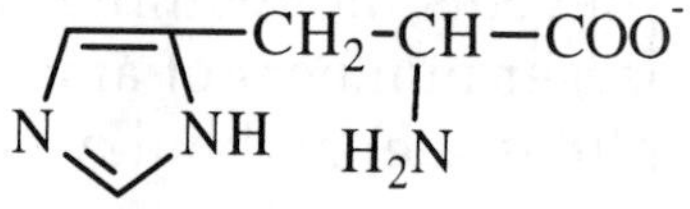

b. The pK$_a$ values for each species are approximately 1.8, 6 and 9.2, respectively.

c. The isoelectric point for histidine is $(6 + 9.2)/2 = 7.6$

6. The resonance forms of the peptide bond in glycylglycine are as follows:

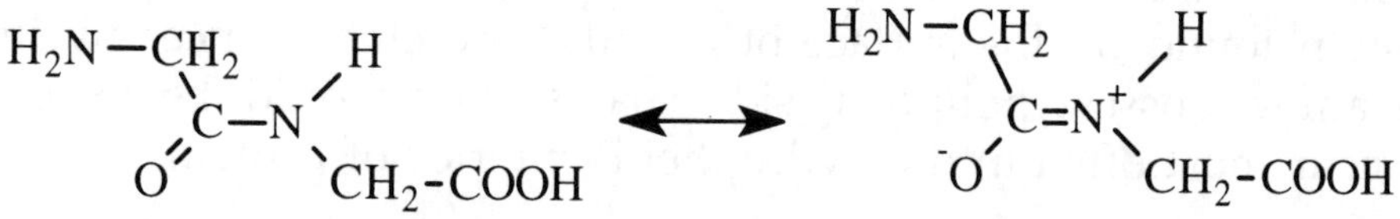

$$H_2N-CH_2-\underset{O}{\overset{}{C}}-N-\underset{}{\overset{H}{}}CH_2-COOH \longleftrightarrow H_2N-CH_2-\underset{O^-}{\overset{}{C}}=N^+-\underset{}{\overset{H}{}}CH_2-COOH$$

The partial double bond character of the peptide bond makes it rigid and planar. Rotation around this bond is therefore hindered.

8. a. Fibrous proteins, which possess water-insoluble sheetlike or ropelike shapes, typically have structural roles in living organisms. Globular proteins are compact spherical molecules (usually water-soluble) that typically have dynamic functions.

   b. Simple proteins contain only amino acids. A conjugated protein is simple protein combined with a nonprotein component, such as lipids or sugars.

   c. An apoprotein is a protein without its prosthetic group. An apoprotein molecule combined with its prosthetic group is referred to as a holoprotein.

10. a. The amino acid sequence is a polypeptide's primary structure.

    b. β-pleated sheet is one type of secondary structure.

    c. Inter- and intra-chain hydrogen bonds between N-H groups and carbonyl groups of peptide bonds are the principal feature of secondary structure. Hydrogen bonds formed between polar side chains are important in tertiary and quaternary structure.

    d. Disulfide bonds are strong covalent bonds that contribute to tertiary and quaternary structure.

12. The structural features of several amino acids do not foster α-helix formation. Because the R group of glycine is too small, the polypeptide chain becomes too flexible. Proline's rigid ring prevents the required rotation of the N-C bond. Sequences with larger numbers of amino acids with charged side chains (e.g., glutamate) and bulky side chains (e.g., tryptophan) are also incompatible with α-helix formation.

14. Amino acids with hydrophobic (nonpolar) R groups are often found in the relatively anhydrous environment of a globular protein's interior because of hydrophobic interactions. Amino acids with polar or charged R groups are often found on protein surfaces because of their capacity to interact with water. Because its side chain possesses a polar amide group, glutamine is expected to be often found on the surface of globular proteins. Since glycine and alanine possess nonpolar side chains they would be expected to occur most often in the hydrophobic interior of proteins.

16. Living cells possess complex mechanisms for assisting the proper folding of nascent polypeptides. These mechanisms are poorly understood and cannot yet be duplicated in the laboratory.

18. The immobilized water of protein molecules is locked into position by hydrogen bonding between polar and ionic groups and water molecules. This gives rise to a three-dimensional structure in which the water molecules have very little freedom of motion, i.e., they are frozen in place.

20. The hydrophobic amino acid side chains are excluded from the water and tend to cluster together. This clustering holds portions of the polypeptides in a particular conformation.

22. The first step in the isolation of a specific protein is the development of an assay which allows the investigator to detect it during the purification protocol. Next, the protein, as well as other substances, are released from source tissue by cell disruption and homogenization. Preliminary purification techniques include salting out in which large amounts of salt are used to induce protein precipitation, and dialysis in which salts and other low molecular weight material are removed. Further purification methods, which are adapted to each research effort at the discretion of the investigator, include various types of chromatography and electrophoresis.

24. Three chromatographic methods commonly used in protein research are ion-exchange chromatography (molecules are separated on the basis of charge differences), gel-filtration chromatography (molecules are separated on the basis of size and shape), and affinity chromatography (a specific molecule is separated from impurities because it binds to a specific ligand that is covalently bound to the column matrix).

26. Because of technical limitations, long polypeptides cannot be directly sequenced. Therefore, a polypeptide is broken into smaller peptides. Several experiments, each of which utilizes a different reagent or proteolytic enzyme, are required because the polypeptide's entire sequence is determined from overlapping sets of sequences.

28. The electrically neutral form of the peptide is:

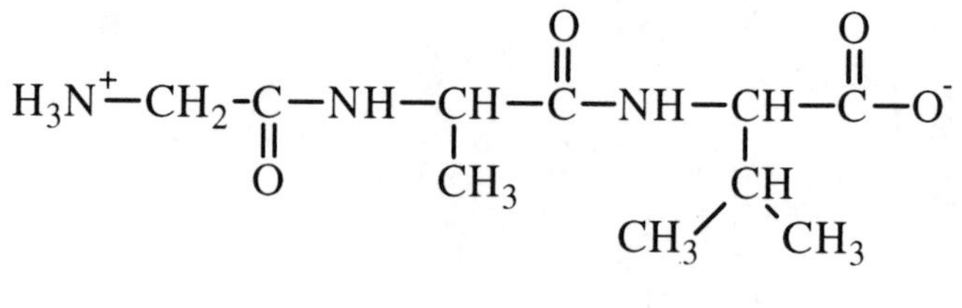

$$H_3\overset{+}{N}-CH_2-\underset{\underset{O}{\|}}{C}-NH-\underset{\underset{CH_3}{|}}{CH}-\overset{\overset{O}{\|}}{C}-NH-\underset{\underset{\underset{CH_3\quad CH_3}{\diagup \diagdown}}{CH}}{CH}-\overset{\overset{O}{\|}}{C}-O^-$$

   a. The approximate value of the isoelectric point is calculated as follows:

   $$pI = (9.6 + 2.32)/2 = 5.96$$

   b. At the pH values of 1 and 5, the peptide will be positively charged and move to the cathode. At the pH values 10 and 12, the peptide will be negatively charged and move to the anode.

30. Proline and hydroxyproline are both imino acids and do not lose their nitrogen atom when they react with ninhydrin. As a result when proline reacts with ninhydrin, the compound shown below is formed.

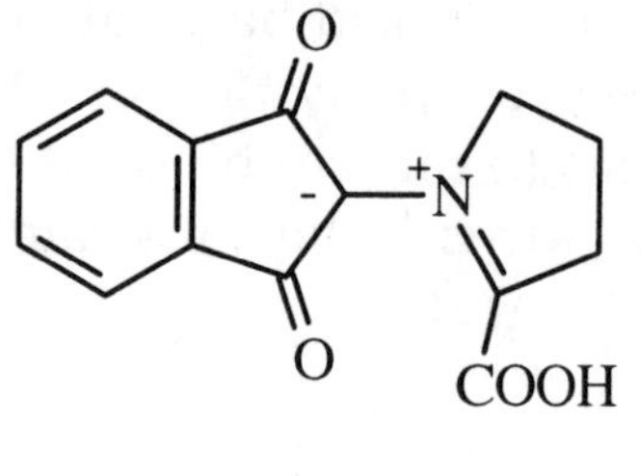

Chapter 6 Answers to Even Numbered Questions

2. The important properties of enzymes are high catalytic rates, a high degree of substrate specificity, negligible formation of side products and capacity for regulation.

4. (a) oxidoreductase, (b) transferase, (c) hydrolyase, (d) lyase, (e) isomerase, (f) ligase.

6.

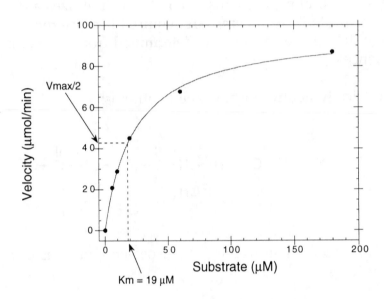

8. Two major types of enzyme inhibitors are competitive and noncompetitive inhibitors. Malonate, whose structure resembles that of succinate, is a competitive inhibitor of succinate dehydrogenase. AMP is a noncompetitive inhibitor of fructose bisphosphate phosphatase.

10. The factors that affect enzyme catalysis include proximity and strain effects (by the correct positioning of substrate in the active site in relation to catalytic groups and the straining of the enzyme-substrate complex as the transition state is achieved), electrostatic effects (the charge distribution in the active site influences the reactivity of the substrate), acid-base catalysis (chemical groups can often be made more reactive by the addition or removal of a proton) and covalent catalysis (a nucleophilic side chain group forms an unstable covalent bond with the substrate).

12. Negative feedback inhibition is a type of allosteric regulation in which the product of a biochemical pathway inhibits the activity of a pacemaker enzyme. The product molecule acts as a negative effector.

14. The major recognized coenzymes are thiamine pyrophosphate (decarboxylation); FAD, FMN, NAD^+, and $NADP^+$ (redox reactions); pyridoxal phosphate (amino group transfer); coenzyme A (acyl transfer reactions); biocytin (carboxylation reactions); tetrahydrofolate (one-carbon group transfer reactions); and deoxyadenosylcobalamin and methylcobalamin (intramolecular rearrangements).

16. The mechanism of chymotrypsin (an enzyme that catalyzes the hydrolysis of peptide bonds) is as follows: The nucleophilic hydroxyl oxygen of Ser-195 launches a nucleophilic attack on the carbonyl carbon of the substrate. The oxyanion that forms as negative charge moves to the carbonyl oxygen is stabilized by hydrogen bonds to the amide NH of Gly-192 and Ser-195. The tetrahedral intermediate formed during this process then decomposes to form the covalently bound acyl-enzyme intermediate. The process then reverses. With water acting as a nucleophile a tetrahedral (oxyanion) intermediate is formed. Subsequently, the bond between the serine oxygen and the carbonyl carbon breaks and the product is released.

18. Enzymes decrease the activation energy required for a chemical reaction because they provide an alternate reaction pathway that requires less energy than the uncatalyzed reaction. They do so principally because of the unique, intricately shaped active sites which possess strategically placed amino acid side chains, cofactors and coenzymes that actively participate in the catalytic process.

20. The pK_a of the imidazole group of histidine is approximately 6. Therefore, the histidine side chain ionizes within the physiological pH range. The protonated form of histidine is a general acid, and the unprotonated form is a general base.

22. Enzymes are composed of amino acids, most of which are optically active. Recall that L-isomers of amino acids are usually found in proteins. Therefore, the active sites of enzymes are also asymmetric, binding one stereoisomeric form of the substrate.

Chapter 7 Answers to Even Numbered Questions

2.   a.   Glucose and mannose are examples of epimers.
       b.

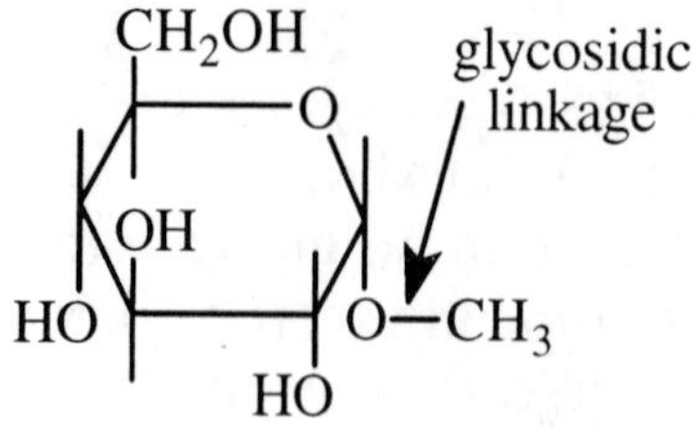

       c.   Glucose is a reducing sugar.
       d.   Ribose is a monosaccharide.
       e. α- and β-Glucose are anomers.
       f   Glucose and galactose are diastereomers.

4.   a.   Ribonuclease B is an example of a glycoprotein.
       b.   Each proteoglycan contains glycosaminoglycans such as chondroitin sulfate and dermatan sulfate which are linked to a core protein via glycosidic linkages.
       c.   Lactose is an example of a disaccharide.
       d.   Heparin is a glycosaminoglycan.

6.

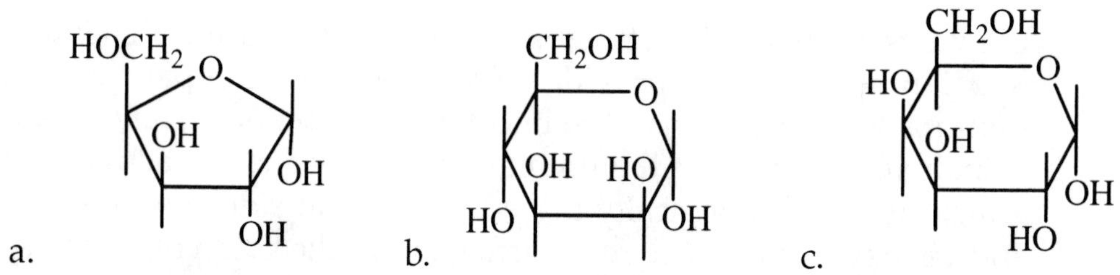

a.   b.   c.

8.   Starch and glycogen are both homopolysaccharides containing glucose monomers linked by α (1,4) glycosidic bonds with branch points connected by α (1,6) glycosidic bonds. Glycogen, however, is much more highly branched than starch. Cellulose is a linear polymer of glucose linked by β (1,4) glycosidic bonds.

10.   When steroids are conjugated with a uronic acid, the OH groups of the uronic acid hydrogen bonds with the water. This structural feature increases the solubility of the conjugated steroid molecule.

12.   Carbohydrates linked by α (1,4) glycosidic bonds without branch points are usually coiled into a helix.

14. a. The three monosaccharide units of raffinose are galactose, glucose, and fructose.
    b. The linkage between the galactose and glucose is α; the linkage between the glucose and fructose is β.
    c. Raffinose is a nonreducing sugar.
    d. Raffinose is not capable of mutarotation.

16. The OH group is not easily displaced in nucleophilic substitution reactions. The phosphate group, on the other hand, is a good leaving group and easily displaced. The formation of the phosphate ester increases the reactivity of that particular site on the carbohydrate molecule.

18.

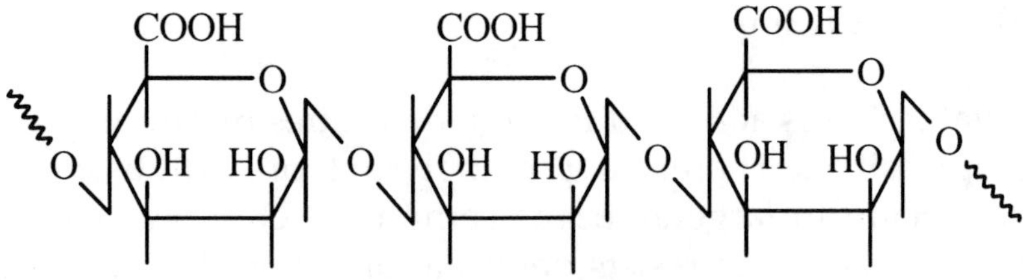

Alginic Acid

The polymer acts to immobilize water through extensive hydrogen bonding.

20. In glycoproteins carbohydrate moieties are most frequently linked to asparagine, serine and threonine.

22.

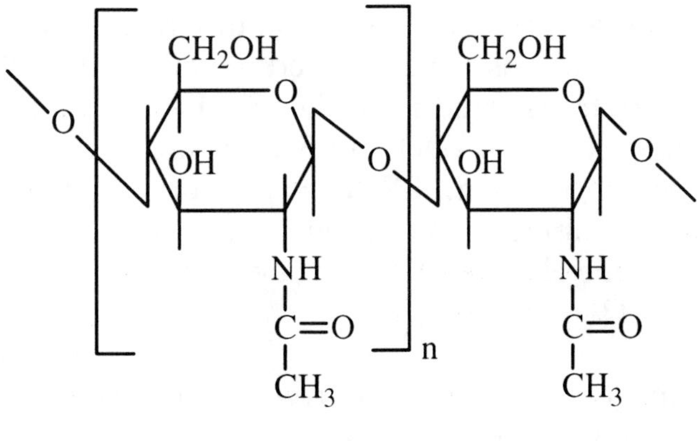

Chitin

189

24. Proteoglycans are distinguished from glycoproteins by their extremely high carbohydrate content and the repeating disaccharide structure of GAG chains. The proteoglycans, which contain a large number of sugar molecules, bind large volumes of water and can contribute support and elasticity to tissues. The carbohydrate components of glycoproteins vary from mono- and disaccharides to branched oligosaccharides. The functions of carbohydrate in individual glycoprotein molecules include protection from denaturation to biological functions such as recognition phenomena (e.g., cell-cell and cell-virus interactions).

26. In cellulose the glucose residues are linked by β (1,4) glycosidic bonds unlike starch and glycogen which possess α (1,4) links. In contrast to amylopectin and glycogen, cellulose does not possess α (1, 6) glycosidic linkages.

28. The water that is absorbed in large quantities by proteoglycans is incompressible. Therefore, proteoglycans provide tissues that contain them in large amounts some protection against mechanical stress, i.e., the tissue resists deformation when pressure is applied.

Chapter 8 Answers to Even Numbered Questions

2. Catabolism consists of three stages. In stage 1, the major nutrient molecules (protein, carbohydrates and fats) are degraded to building block molecules in a process called digestion. In stage 2, amino acids, monosaccharides, glycerol and fatty acids produced in stage 1 are converted to a smaller group of molecules. The most important product of stage 2 is acetyl-CoA. During stage 3, the acetyl group of acetyl-CoA enters the citric acid cycle in which it is completely oxidized. The carbons are converted to carbon dioxide and the hydrogens are transferred to NAD^+ and FAD. A large amount of energy is released as NADH and $FADH_2$ are re-oxidized in the electron transport system. Water is produced as two electrons and two protons combine with oxygen. A portion of the energy generated by the electron transport system is captured and used to synthesize ATP.

4. The ATP and NADPH used to drive numerous anabolic processes are produced by catabolic reactions.

6. In an enzyme cascade, a series of enzymes that are present in their precursor form are sequentially converted to their active forms. The process is often initiated by the binding of a second messenger to a specific proenzyme that results in its conversion to an active

enzyme. The enzyme then catalyzes the activation of a second enzyme (enzyme b). Because each initial enzyme (enzyme a) converts numerous copies of proenzyme b to active enzyme b the signal is rapidly amplified.

8. a. Insulin stimulates glycogenesis and inhibits glycogenolysis.
    b. Glucagon stimulates glycogenolysis and inhibits glycogenesis.
    c. Fructose-2,6-bisphosphate is an effector molecule that activates PFK- 1 and stimulates glycolysis.

10. In the synthesis of new glycogen molecules, a primer protein called glycogenin is used to initiate glycogen formation. Glucose is transferred from UDP-glucose to a specific tyrosine residue of the glycogenin. This glucose then serves as the starting point for a new growing glycogen molecule.

12. Glycolysis occurs in two stages. In stage I, glucose is phosphorylated and cleaved to two molecules of glyceraldehyde-3-phosphate. During this stage two, ATP molecules are consumed. In stage 2, two molecules of glyceraldehyde-3-phosphate are converted to pyruvate, a process in which four ATP and two NADH are produced.

14. Phosphoenolpyruvate has a high phosphate group transfer potential because the transfer of the enol phosphate to another molecule produces a vinyl alcohol. The vinyl alcohol tautomerizes rapidly to the keto-form making the transfer almost irreversible.

16. If gluconeogenesis and glycolysis were exactly the reverse of one another, futile cycles would be established and much energy would be wasted. In addition, it would be impossible for the body to store glycogen or release glucose into the blood as needed.

18. Futile cycles are prevented by having the forward and reverse reactions catalyzed by different enzymes, both of which are independently regulated.

Chapter 9 Answers to Even Numbered Questions

2. a. Phospholipids perform major roles as structural components of membranes, emulsifiers and surface active agents.
    b. Plant and animal membranes contain large amounts of sphingolipids.
    c. Oils serve as an important energy reserve of fruits and seeds.
    d. Waxes serve as protective coatings on the surface of leaves and stems, on the fur of animals and on the shells of insects.

e.   Steroids play an important structural role in animal membranes. Certain steroids act as hormones.

f.   By acting as light trapping pigments, carotenoids play an important role in photosynthesis.

4.   Membranes with high levels of unsaturated fatty acids tend to be more fluid and have lower freezing points. Such membranes can play an important role in animal species exposed to intense cold. Because of their proximity to cold surfaces, hooves and legs chill easily. Membranes with a high unsaturated lipid content and a low freezing point could maintain their function under these extreme conditions.

6.   The polar head region containing the ester and amide linkages would be hydrophilic; the hydrocarbon tail region would be hydrophobic.

8.   Plasma lipoproteins transport lipid molecules from one organ to another through the bloodstream. The protein component serves to solubilize the complexes in the blood.

10. Unsaturated fatty acid content increases fluidity, whereas cholesterol decreases fluidity.

12. Both a and c are true. (Ionophores are discussed on page 310.)

14. Lateral movement occurs easily in membranes because hydrophobic and hydrophilic groups maintain their relative positions and there is no change in the types of noncovalent bonding involved in the membrane structure. Trans membrane movement is more difficult because hydrophilic groups must move through a hydrophobic region in the membrane bilayer. This process requires the input of additional energy and is, therefore, significantly slower.

16. When acetylcholine binds to the acetylcholine receptor complex in muscle cell plasma membrane, sodium ions flow into the muscle cell and a smaller number of potassium ions flow out. During the repolarization phase of muscle contraction, potassium ions flow out of the cell through voltage-regulated potassium channels.

18. Prostaglandins have major recognized roles in all of the following processes: reproduction, digestion, respiration, inflammation, and smooth muscle contraction.

20. Membrane structure is not a function of triacylglycerols. Membrane structure is a function of phospholipids.

22. Lipids are not directly involved in the active transport of ions across membranes. Lipids are responsible for the following properties of membranes: selective permeability, self-sealing capability, fluidity and membrane asymmetry.

24. Each type of transport protein or carrier binds a specific molecule. As the result of this binding, a conformational change in the carrier occurs, thereby causing translocation across the membrane. The red blood cell glucose transporter is an example of such a carrier.

26. As time progresses, the antigens of the heterokaryon will intermingle. This suggests that the membrane is fluid and the antigens, as well as other components of the cell membrane, can move freely within the lipid bilayer.

Chapter 10 Answers to Even Numbered Questions

2.
   a. Carnitine is an amino acid that is used to transport acyl-CoA molecules into the mitochondria.
   b. Flippase is a protein that transfers choline-containing phospholipids across the membrane.
   c. Thrombin is a proteolytic enzyme that converts fibrinogen to fibrin.
   d. Thiolase catalyzes the final reaction in the β-oxidation cycle, referred to as a thiolytic cleavage, which yields an acetyl-CoA product.
   e. Desmolase catalyzes the initial reaction in steroid hormone synthesis, the conversion of cholesterol to pregnenolone.

4. Aspirin inhibits the synthesis of prostaglandins released during small injuries associated with exercise. Inhibition of cyclooxygenase prevents the formation of TXA_2, a molecule produced in platelets that is involved in the clotting of blood.

6. Fatty acid synthesis in plants differs from that in animals in the following ways: location (plant fatty acid synthesis occurs mainly in the chloroplasts, whereas in animals fatty acid biosynthesis occurs in the cytoplasm), metabolic control (in animals the rate limiting step is catalyzed by acetyl-CoA carboxylase, whereas in plants, this does not appear to be the case), enzyme structure (the structure of plant acetyl-CoA carboxylase and fatty acid synthetase are more closely related to similar enzymes in *E. coli* than to those in animals).

8. Steroids are compounds containing the following ring system:

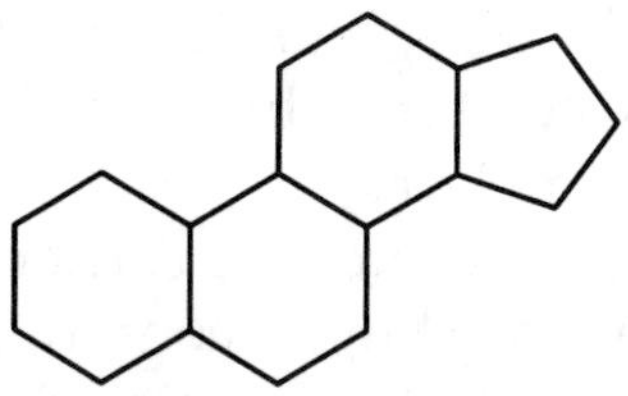

The term steroid is often used to designate all compounds containing this ring system. It is more accurately reserved for those derivatives that contain carbonyl groups. Sterols have a similar structure but also contain hydroxyl groups.

10. Conjugation increases the number of groups in the molecule that are capable of hydrogen bonding with water. The greater the degree of hydrogen bonding with water, the greater the solubility of the molecule.

12. In periods of fasting blood glucose levels fall and glucagon and epinephrine are released. These hormones then bind to their respective adipocyte plasma membrane receptors. This binding initiates a cascade (cAMP activates protein kinase which in turn activates hormone sensitive lipase) that results in the release of fatty acids and glycerol into the blood.

14. Insulin facilitates the transport of glucose into the adipocytes and stimulates fatty acid synthesis, thereby promoting triacylglycerol synthesis. It also prevents lipolysis by inhibiting protein kinase.

16. Although regular eating is not a panacea, it does provide sufficient carbohydrate to act as a fuel to sustain vital metabolic processes.

18. The two long chain fatty acids attached through ester linkages to C-1 and C-2 of glycerol are hydrophobic. The polar head consisting of a negatively charged phosphate group and a positively charged quaternary nitrogen group attached to C-3 is hydrophilic. This molecule would insert into membrane with its polar head on the surface and the two fatty acid components in the hydrophobic core.

20. Severe dieting significantly reduces caloric intake thereby stimulating the generation of energy by massive lipolysis. The large amounts of acetyl-CoA that results from this process cause the synthesis of similarly large amounts of ketone bodies. When present in significant amounts, ketone bodies overwhelm the blood's buffering capacity and the pH falls.

Chapter 11 Answers to Even Numbered Questions

2. Ancient earth possessed a reducing atmosphere that mainly consisted of ammonia and methane. With the development of photosynthesis, oxygen was released into the atmosphere. Subsequently, this oxygen reacted with methane to form carbon dioxide and with ammonia to form molecular nitrogen. The continued release of oxygen produced an oxidizing atmosphere consisting of primarily oxygen, nitrogen and carbon dioxide.

4. Contracting muscle converts large amounts of ATP to ADP in the muscle cells. The drop in ATP concentration stimulates two key regulatory enzymes of the citric acid cycle. (1) Citrate synthetase catalyzes the condensation of acetyl-CoA and oxaloacetate to give citrate. ATP is an allosteric inhibitor of this enzyme. As concentrations of ATP drop, this enzyme becomes more active. (2) Isocitrate dehydrogenase, which converts isocitrate to α-ketoglutarate is inhibited by high concentrations of ATP and activated by high concentrations of ADP. A reduced ATP concentration also stimulates the conversion of pyruvate to acetyl-CoA catalyzed by pyruvate dehydrogenase.

6. Citrate, produced in the mitochondria by the citric acid cycle, is transported across the mitochondrial membrane into the cytoplasm. Once in the cytoplasm, citrate is cleaved by citrate lyase to acetyl-CoA and oxaloacetate. The acetyl-CoA is then used to synthesize fatty acids as well as other biomolecules. The oxaloacetate is reduced to malate and moved across the mitochondrial membrane where it is reoxidized to oxaloacetate, the citric acid cycle intermediate.

8. A high NADH/NAD⁺ ratio (a) indicates that a cell's energy requirements are currently being met. NADH inhibits pyruvate dehydrogenase, citrate synthase, isocitrate dehydrogenase and α-ketoglutarate dehydrogenase. A high ADP/ATP ratio (b) and low citrate concentration (d) are indicators of a low cell energy state. A high ADP/ATP ratio (b) stimulates isocitrate dehydrogenase, citrate synthase, ATP synthase and oxidative phosphorylation. A high acetyl-CoA concentration (c) inhibits pyruvate dehydrogenase and stimulates pyruvate carboxylase and fatty acid synthesis and stimulates PFK-1. A high succinyl-CoA concentration (e) inhibits citrate synthase and α-ketoglutarate dehydrogenase.

10. Principal sources of electrons for the mitochondrial electron transport system are NADH and $FADH_2$.

12. The chemiosmotic coupling theory has the following principal features: As electrons pass through the electron transport chain (ETC), protons are transported from the matrix and released into the inner membrane space. As a result, an electrical potential (Ψ) and a proton gradient (ΔpH) are created across the inner membrane. The electrochemical proton gradient is sometimes referred to as the proton motive force (Δp). Protons, which are present in the inter-membrane space in great excess, can only pass down their concentration gradient, across the inner membrane, and back into the matrix, through the proton translocating ATP synthase.

14. ATP synthesis requires the presence of a proton gradient across the mitochondrial membrane. Dinitrophenol, which has an ionizable hydrogen, can diffuse across the mitochondrial membrane. As it diffuses across the membrane, protons are transported from one side of the mitochondrial membrane to the other, thereby disrupting the proton gradient and interfering with ATP synthesis.

16. The pentose phosphate pathway provides NADPH for reductive reactions such as lipid biosynthesis and antioxidant mechanisms, and ribose-5-phosphate (a molecule used in nucleotide and nucleic acid biosynthesis). In plants, the pentose phosphate pathway is involved in the synthesis of glucose during the light-independent reactions of photosynthesis.

18. Examples c, d and f are all reactive oxygen species. Each of these species can act as a free radical, i.e., it can attack various cell components producing such effects as enzyme inactivation, polysaccharide depolymerization, DNA breakage and membrane destruction.

20. The major enzymatic defenses against oxidative stress are provided by superoxide dismutase, catalase and glutathione peroxidase. Superoxide dismutase converts the superoxide radical to hydrogen peroxide. Catalase uses H_2O_2 to oxidize substrates and converts excess H_2O_2 to H_2O and O_2. Glutathione peroxidase uses the reducing agent glutathione (GSH) to convert H_2O_2 to H_2O and transforms organic peroxides to alcohols.

22. Complex reaction pathways have many control points, thereby making possible the delicate manipulation of metabolism in living cells which are exposed to complex environmental conditions.

24. Because of the possibility of isomeric structures being produced either in the formation or hydration of a double bond, after one turn of the citric acid cycle, every carbon of the oxaloacetate would have some radiolabel.

Chapter 12 Answers to Even Numbered Questions

2. The most significant contribution of early photosynthesizing organisms to the earth's environment was the conversion of a reducing atmosphere (ammonia and methane) to an oxidizing atmosphere.

4. Chloroplasts resemble mitochondria in the following ways: (1) they are both similar in size and structure to modern prokaryotes; (2) they both reproduce by binary fission; (3) the genetic information and protein synthesizing capability of both chloroplasts and mitochondria are similar to that of prokaryotes; (4) the ribosomes of chloroplasts and mitochondria are similar in size and function; and (5) they are both thought to have arisen from ancient free living prokaryotes.

6. The final electron acceptor in photosynthesis is carbon dioxide.

8. The net production of the dark, or light-independent, reactions of photosynthesis is the production of one molecule of glyceraldehyde-3-phosphate. See page 347 for the reactions of the Calvin cycle.

10. Conjugation is a system of alternate double and single bonds. When light with sufficient energy strikes the it electrons of a conjugated system (or any double bond), an electron is promoted from the ground state to a higher energy state, referred to as the excited state. Conjugation lowers the energy difference between the ground state and the excited state, hence photons of lower energy are capable of achieving this transition.

12. Only light of a particular energy can be absorbed in photosynthesis. Increasing the intensity increases the number of photons present and hence improves the rate of photosynthesis. Increasing the energy level of the light does increase the energy of the photons. Increasing the energy of the photons may actually decrease the rate of photosynthesis by shifting the photons to an energy not absorbed by the photosynthetic system.

14. The Z scheme is a mechanism whereby electrons are transferred from water to $NADP^+$. This process produces the reducing agent NADPH required for fixing carbon dioxide in the light-independent reactions of photosynthesis. Removal of the electrons from water also results in the production of oxygen. As electrons flow from PSII to PSI, protons are pumped across the thylakoid membrane, a process that establishes the proton gradient that drives ATP synthesis.

16. Carbon dioxide fixation occurs in the stroma of the chloroplasts.

18. If sufficient carbon dioxide is already present to saturate all of the ribulose- 1,5-bisphosphate carboxylase molecules, the presence of additional carbon dioxide molecules will not increase the rate of photosynthesis. In addition, photosynthesis is depressed by low light levels.

20. Photorespiration is carried out by RuBisCO, an enzyme that has both oxygenase and carbon dioxide fixation activities. Under conditions of high carbon dioxide concentration, the oxygenase functions are repressed and photorespiration slows down.

22. Because of the capacity of C4 plants to avoid the process of photorespiration, herbicides that promote photorespiration do not affect these organisms.

24. The maximum rate of photosynthesis will occur at the λ_{max} of the photosynthesizing system. This absorption maximum should match the absorption maximum of the light absorbing pigments. (This topic is described in Appendix B, supplement 12)

26. If blue wavelengths are used in addition to red ones, the rate of oxygen evolution is increased. This phenomenon is known as the Emerson enhancement effect. If photosynthesis occurs in a single photosystem, then the magnitude of the enhancement should reflect the ratios of the λ_{max} in the red and blue regions. The enhancement was much greater than predicted by this ratio. The existence of a second chromophore (and by inference a second photosystem) was therefore suggested.

Chapter 13 Answers to Even Numbered Questions:

2.   Transamination reactions represent a method of conserving valuable nitrogen reserves. The reaction is reversible and can be used to convert α-keto acids produced by metabolic reactions to α-amino acids that may be in short supply. Surplus amino acids present in larger amounts than required by current metabolic needs are used as the source of the amino groups.

4.   Nitrogen-fixing organisms solve the problem of oxygen inactivation in several ways. These are: (1) anaerobic organisms live only in anaerobic soil and are not faced with the problem of oxygen inactivation, (2) other organisms physically separate oxygen from the nitrogenase complex. For example many of the cyanobacteria produce specialized nitrogenase-containing cells called heterocysts. The thick cell walls of the heterocysts isolate the enzymes from atmospheric oxygen. In addition, legumes produce an oxygen binding protein called leghemoglobin which traps oxygen before it can interact with the nitrogenase complex.

6.   The Schiff base formed when pyridoxal phosphate reacts with an amino acid loses a proton to form a carbanion. The free electrons of the carbanion are in conjugation with the positively charged nitrogen of the pyridinium ion. As electrons flow to the positively charged nitrogen, the double bond system reorganizes itself to quench the charge on the nitrogen.

8.   The amino acids whose blood levels are not affected by passage through the liver are b (isoleucine) and d (valine).

10.   a.   Glutamine is produced in the following reactions:

        α-Ketoglutarate + NADH + NH_3 + H^+ $\rightarrow$ Glutamate

        Glutamate + NH_3 + ATP $\rightarrow$ Glutamine + ADP + P_i

    b.   Methionine is produced in the following series of reactions:

        L-Aspartate + ATP $\rightarrow$ β-Aspartylphosphate + P_i + ADP

        β-Aspartylphosphate + NADPH + H^+ $\rightarrow$ Aspartate β-semialdehyde

        Aspartate β-semialdehyde + NADPH + H^+ $\rightarrow$ Homoserine

        Homoserine $\rightarrow$ Cysteine $\rightarrow$ Homocysteine $\rightarrow$ Methionine

    c.   Homoserine produced as in part b above reacts as follows:

        Homoserine + ATP $\rightarrow$ Phosphohomoserine + ADP + H^+

        Phosphohomoserine + H_2O $\rightarrow$ Threonine + P_i

12. a. Alanine belongs to the pyruvate family.
     b. Phenylalanine belongs to the aromatic family.
     c. Methionine belongs to the serine family.
     d. Tryptophan belongs to the aromatic family.
     e. Histidine belongs to the histidine family.
     f Serine belongs to the serine family.

14. Glutathione is involved in the synthesis of DNA, RNA and the eicosanoids. It is also utilized as a reducing agent which protects cells from radiation and oxygen, and a conjugating agent for environmental toxins. Glutathione is also believed to play a role in amino acid transport.

16. a. Nucleotide, b. Nucleoside, c. Purine, d. Pyrimidine, and e. Nucleotide (nucleoside triphosphate)

18. Pyrimidine nucleosides occur predominantly in the anti conformation. Steric hindrance between the pentose sugar and the carbonyl oxygen at C-2 of the pyrimidine ring prevents free rotation around the N-glycosidic bond.

20. Five ATP are required to synthesize a purine by the *de novo* pathway. Only one ATP is required if a purine molecule is recovered by the salvage pathway thus giving a total of four ATP.

22. The 10 essential amino acids for adult humans are: isoleucine, leucine, lysine, methionine, phenylalanine, threonine, tryptophan, and valine. In addition, histidine and arginine are essential for infants. Essential amino acids cannot be synthesized by the body and must therefore be consumed in the diet.

24. In the γ-glutamyl cycle glutathione (GSH) is excreted from the cell. γ-Glutamyltranspeptidase converts GSH to a γ-glutamylamino acid derivative and Cys-Gly. The γ-glutamylamino acid derivative is transported into the cell where it is converted to 5-oxoproline and the free amino acid. 5-Oxoproline is eventually reconverted to GSH. The Cys-Gly is transported into the cell where it is hydrolyzed to cysteine and glycine. The location of the γ-glutamyltransferase on the plasma membrane facilitates the transport process.

26. Glutamate plays a central role in amino acid metabolism because it and α-ketoglutarate constitute one of the most common α-amino acid/ α-ketoacid pairs used in transamination reactions. Glutamate also serves as a precursor of several amino acids and as a component of polypeptides. Glutamine serves as the amino group

donor in numerous biosynthetic reactions (e.g., purine, pyrimidine and amino sugar synthesis), as a safe storage and transport form of ammonia, and as a component of polypeptides.

28. During the reaction catalyzed by serine hydroxymethyltranferase the radiolabeled carbon atom of serine enters the THF pool as N^5,N^{10}-methylene THF. Because N^5,N^{10}-methylene THF is reversibly converted to N^5,N^{10}- methynyl THF and N^{10}-formyl THF (the coenzyme used in purine synthesis) some radiolabeled carbon atoms will enter the purine synthetic pathway. Because N^{10}-formyl THF is a required coenzyme in the reactions in which 5'-phosphoribosyl-N-formyl-glycinamide and 5-phosphoribosyl-4-carboxamide-5-foramidoimidazole are synthesized, ^{14}C will appear as C-8 and C-2 of the purine ring (shown below).

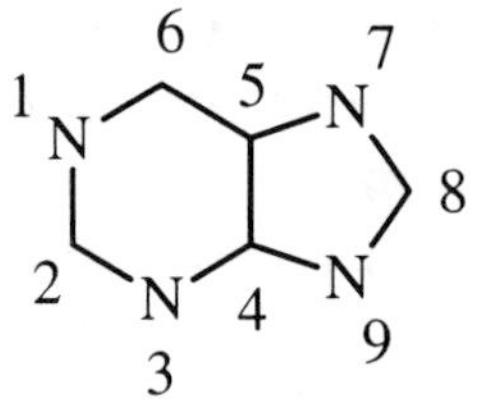

30. In ping-pong reactions, the first substrate must leave the active site before the second can enter. In the reaction of alanine with α-ketoglutarate to produce pyruvate and glutamate the following steps take place: (1) the alanine enters the active site and transfers the amino group to pyridoxal phosphate, (2) water enters the reaction site and hydrolyses the Schiff base to produce pyridoxamine phosphate and pyruvate, (3) pyruvate diffuses from the active site, (4) α-ketoglutarate, the second substrate, enters the reaction site and forms a Schiff base with the pyridoxamine phosphate, (5) water hydrolyses the Schiff base to give pyridoxal phosphate and glutamate, and (6) glutamate diffuses out of the active site.

32. The biologically active form of folic acid referred to as tetrahydrofolate or THF is shown below.

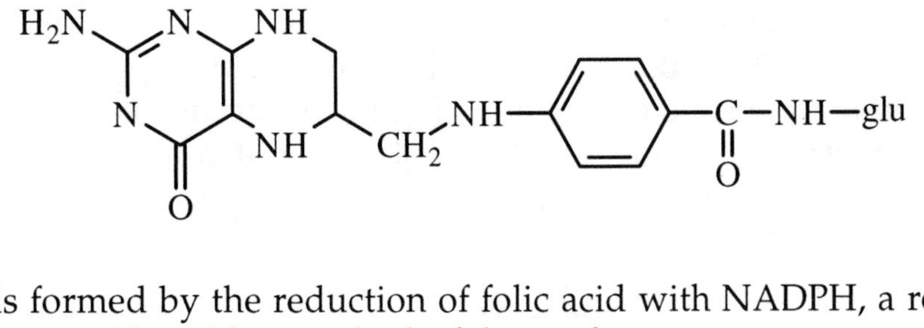

It is formed by the reduction of folic acid with NADPH, a reaction that is catalyzed by tetrahydrofolate reductase.

Chapter 14 Answers to Even Numbered Questions

2. The major nitrogen-containing excretory molecules are ammonia, urea, uric acid, allantoin, and allantoate.

4. The structural features that apparently mark proteins for destruction are: (1) certain N-terminal amino acid residues (e.g., methionine or alanine), (2) Peptide motif sequences (e.g., amino acid sequences with proline, glutamic acid, serine and threonine), and (3) oxidized residues (amino acid residues whose side chains have been oxidized by oxidases or ROS).

6. a. Ketogenic, b. Ketogenic, c. Glycogenic, d. Glycogenic, e. Glycogenic, f. Both.

8. Ammonia can be used to excrete waste nitrogen in certain aquatic species because they live in water. The toxic effects of ammonia are therefore immediately diminished after it is excreted into a large body of water. This strategy is not practical for mammals, most of which are land animals. The toxic effects of ammonia, a relatively powerful base and a strong nucleophile, include depletion of glutamate and α-ketoglutarate, inhibition of amino acid transport and the Na^+-K^+ dependent ATPase, all of which contribute to brain damage.

10. The first two reactions in the biochemical pathway that converts NH_4^+ to urea (i.e., the formation of carbamoyl phosphate and citrulline) occur in the mitochondrial matrix. Subsequent reactions that convert citrulline to ornithine and urea occur in the cytosol. Both citrulline and ornithine are transported across the inner membrane by specific carriers.

12. Individuals with PKU lack phenylalanine hydroxylase (phenylalanine-4-monooxygenase) activity so they cannot synthesize tyrosine from phenylalanine. Tyrosine is therefore an essential amino acid for these patients.

14. The term Krebs bicycle refers to two interlocking cyclic reaction pathways. The aspartate-argininosuccinate shunt of the citric acid cycle is responsible for regenerating the aspartate needed for the urea cycle from fumarate. The molecule which the two cycles have in common is argininosuccinate.

16.   If any of the urea cycle enzymes is missing, ammonia (the nitrogen-containing substrate of the cycle) cannot be metabolized. If any of the enzymes is defective (i.e., it does not catalyze its reaction at an appropriate rate) ammonia is metabolized slowly. Under both circumstances the body's ammonia concentration is excessively high.

18.   5,6,7,8-Tetrahydrobiopterin (BH_4) is required in the synthesis of neurotransmitters such as norepinephrine and serotonin which are produced in brain. BH_4 is required in the reaction catalyzed by tyrosine hydroxylase in which tyrosine is hydroxylated to form L-dopa (a precursor of several neurotransmitters including dopamine and norepinephrine), and the reaction catalyzed by tryptophan hydroxylase in which tryptophan is hydroxylated to form 5-hydroxytryptophan. L-dopa and 5-hydroxytryptophan must be supplied to individuals lacking the capacity to synthesize BH_4 because the latter molecule does not cross the blood-brain barrier.

20.   a. Methylene (CH_2) groups, b. Methyl groups, c. Methyl groups, d. Methyl groups.

22.   Caffeine (trimethylxanthine), which occurs in both beverages, is similar in structure to xanthine, the immediate precursor of uric acid. Some caffeine molecules are excreted as methyluric acid, a molecule with solubility properties similar to those of uric acid. Recall that gout is caused by the precipitation of uric acid in the joints.

24.   In ubiquination, the major mechanism of protein degradation, ubiquitin (a small hsp) is covalently attached to lysine residues of a protein with structural features such as oxidized amino acid residues that mark it for destruction. Once the protein is ubiquinated, it is degraded by proteases in ATP-requiring reactions.

Chapter 15 Answers to Even Numbered Questions

2.   a. Kidney, b. Liver, c. Intestine, d. Brain, e. Adipose tissue, f Liver.

4.   The sustained generation of energy from fat requires a large supply of citric acid cycle intermediates. Recall that oxaloacetate is derived from glucose via the enzymes of the glycolytic pathway and pyruvate carboxylase.

6.	The hypothalamus regulates the function of the pituitary largely by secreting small amounts of specific releasing factors into a specialized capillary bed which directly connects the two structures. If the pituitary were to be transplanted to another part of the body, the concentration of the hypothalamic releasing factors in the blood that reaches the pituitary cells would be too low to affect their function.

8.	NADPH, which is formed during the pentose phosphate pathway and reactions catalyzed by isocitrate dehydrogenase and malic enzyme, is used as a reducing agent in a wide variety of synthetic reactions (e.g., amino acids, fatty acids, sphingolipids and cholesterol). The degradation of some of these molecules (e.g., fatty acids and the carbon skeletons of the amino acids) results in the synthesis of NADH, a major source of cellular energy via the mitochondrial electron transport system.

10.	The recognition by the conscious centers in the brain that danger is imminent results immediately in rapid mobilization of the body's resources. One important consequence of this process is a discharge of epinephrine from the sympathetic nervous system and the adrenal medulla. The large amounts of glucose and fatty acids that flood into blood as a consequence of epinephrine's stimulation of glycogenolysis and lipolysis have several effects on the body. One effect, high blood glucose levels, provides the energy required for rapid decision-making processes in the brain. In addition, large quantities of glucose and fatty acids are required for strenuous physical activity if a decision is made to run away from the danger.

12.	In uncontrolled diabetes mellitus, massive breakdown of fat reserves results in the production of large quantities of ketone bodies. Two of the ketone bodies (acetoacetic acid and β-hydroxybutyric acid) are weak acids. The release of hydrogen ions from large numbers of these molecules overwhelms the body's buffering capacity.

14.	The storage of preformed hormone molecules in secretory vesicles allows for a rapid response of the producing cells to metabolic signals. As soon as the appropriate signal is received the vesicles fuse with plasma membrane and (via exocytosis) release their contents into the blood stream.

16.	The effects of hormone action include changes in gene expression (e.g., increased or decreased synthesis of specific proteins, including regulatory enzymes), and activation or inactivation of preexisting enzyme molecules.

18. Five important types of plant hormones include the auxins (a group of plant growth regulators such as indole acetic acid that increases nucleic acid and protein synthesis), the gibberellins (a group of tetracyclic diterpenes known for their effects in stem growth as well as other processes such as flowering and seed production), the cytokinins (molecules similar in structure to adenine that promote growth and developmental processes such as chloroplast development), abscisic acid (a molecule that antagonizes many growth processes) and ethylene (a molecule that is involved in longitudinal growth, flower senescence and fruit ripening).

20. For several weeks after the onset of fasting, blood glucose levels are maintained via gluconeogenesis. During most of this period amino acids derived from the breakdown of muscle proteins are the major substrates for this process. Eventually, as muscle becomes depleted the brain switches to ketone bodies as an energy source. Consequently, the production of urea (the molecule used to dispose of the amino groups of the amino acids) declines.

22. One consequence of physical activity is the activation of the sympathetic nervous system which in turn stimulates the adrenal gland to secrete epinephrine and norepinephrine. These hormones then activate the adipocyte enzyme hormone-sensitive lipase which catalyzes the hydrolysis of triacylglycerol molecules to form glycerol and the fatty acids used to drive muscle contraction.

24. The capacity to switch from using glucose as an energy source, after gluconeogenic substrates have become scarce, to ketone bodies during starvation, has survival value because it increases the survival time thereby increasing the chances that the organism will eventually find a food source.

26. During the initial phase of a prolonged fast, blood glucose and insulin levels fall, and glucagon release is triggered. Glucagon acts to prevent hypoglycemia by promoting glycogenolysis and gluconeogenesis. The amino acids derived from muscle protein are a major source of the carbon skeleton substrates in gluconeogenesis.

28. HbA_{1c} formation is a consequence of nonenzymatic glycosylation of hemoglobin that occurs in the presence of high blood glucose levels. In the Maillard reaction the aldehyde group of glucose condenses with a free amino group in a protein to form a Schiff base. The Schiff base rearranges to form a stable ketoamine referred to as the Amadori product. The Amadori product subsequently destabilizes to form a reactive carbonyl-containing product that reacts with hemoglobin molecules to form an adduct such as HbA_{1c}.

Chapter 16 Answers to Even Numbered Questions

2. RNA hairpins and DNA:RNA hybrids adopt an A-DNA-like conformation because the 2'-OH of the ribose residue in RNA sterically hinders the adoption of a B-DNA-like helix. Recall that in A-DNA, base pairs tilt 20° away from the horizontal.

4. B-DNA, the right handed double helical structure discovered by Watson and Crick, in which there are 10.4 base pairs per helical turn (3.4 nm; diameter = 2.4 nm), occurs under humid conditions. A-DNA, a partially dehydrated molecule, possesses 11 base pairs per helical turn (2.5 nm; diameter 2.6 nm). The Z form of DNA is twisted into a left handed spiral with 12 base pairs per helical turn. Each helical turn occurs in 4.5 nm with a diameter of 1.8 nm.

6. Biological processes that are facilitated by supercoiling include packaging of DNA into compact forms (i.e., chromosomes), DNA replication and transcription.

8. Eukaryotic genomes are larger than those of prokaryotes. In contrast to prokaryotic genomes, which consist entirely of genes, the majority of eukaryotic DNA sequences do not appear to have coding functions. Most eukaryotic genes are not continuous (i.e., they usually contain introns) unlike those of prokaryotes.

10. RNA molecules differ from DNA in the following ways: (1) RNA contains ribose instead of deoxyribose, (2) the nitrogenous bases in RNA differ from those of DNA (e.g., uracil replaces thymine and several RNA bases are chemically modified), and (3) in contrast to the double helix of DNA, RNA is single-stranded.

12. The three major types of RNA are ribosomal RNA (a component of ribosomes), transfer RNA (each molecule transports a specific amino acid to the ribosome for assembly into proteins), and messenger RNA (each molecule specifies the sequence of amino acids in a polypeptide).

14. First, fractions are recovered from the centrifuge tubes (Refer to Appendix B, supplement 2.1). This is followed by scanning each fraction with UV light in a spectrophotometer. The DNA-containing fractions are easily identified because DNA absorbs strongly at 260 nm. The fractions containing supercoiled and relaxed DNA are distinguished on the basis of density. Because of its compact shape, supercoiled DNA is more dense than relaxed DNA.

16. There are approximately 6 million base pairs in a single human cell. Assuming that there are 10^{14} body cells, the total length of the DNA in the human body is approximately 2×10^{11} km. This estimate is about 1000 times greater than the distance from the earth to the sun.

18. In a relaxed circular DNA molecule the number of helical turns (T or twist) is equal to the linking number. In this example, T is equal to 48 (i.e., 500/10.4). Therefore, before the enzymatic alteration L=48. After the alteration, L=44.

20. One of the principal reasons for the problems in the development of an AIDS vaccine is that the HIV genome mutates frequently. Consequently, the surface antigens of HIV also often change. Note that the antigens in a vaccine stimulate the immune system to produce antibodies that will bind specifically to antigens on the surface of disease-causing organisms.

22. Guanine-cytosine pairs contain three hydrogen bonds while adenine-thymine contain only two. The more hydrogen bonds holding the DNA strands together, the higher the melting point. will be.

Chapter 17 Answers to Even Numbered Questions

2. Negative supercoiling facilitates the unzipping of DNA during the initiation phase of replication.

4. In the Messelson-Stahl experiment all of the nitrogen was ^{15}N. To accomplish the same effect with a carbon isotope, all of the carbon in the medium would have to be isotopically pure in both phases of the experiment. This would be prohibitively expensive.

6. a. Helicase is an enzymatic activity required during DNA uncoiling.
   b. Primase is an enzymatic activity that catalyzes the synthesis of RNA primers.
   c. DNA polymerase is an enzymatic activity that catalyzes several reactions during DNA replication.
   d. DNA ligase forms phosphodiester linkages between newly synthesized DNA fragments.
   e. Topoisomerase is an enzymatic activity that prevent the tangling of DNA strands during DNA replication.
   f DNA gyrase facilitates the separation of DNA strands during prokaryotic replication.

8.

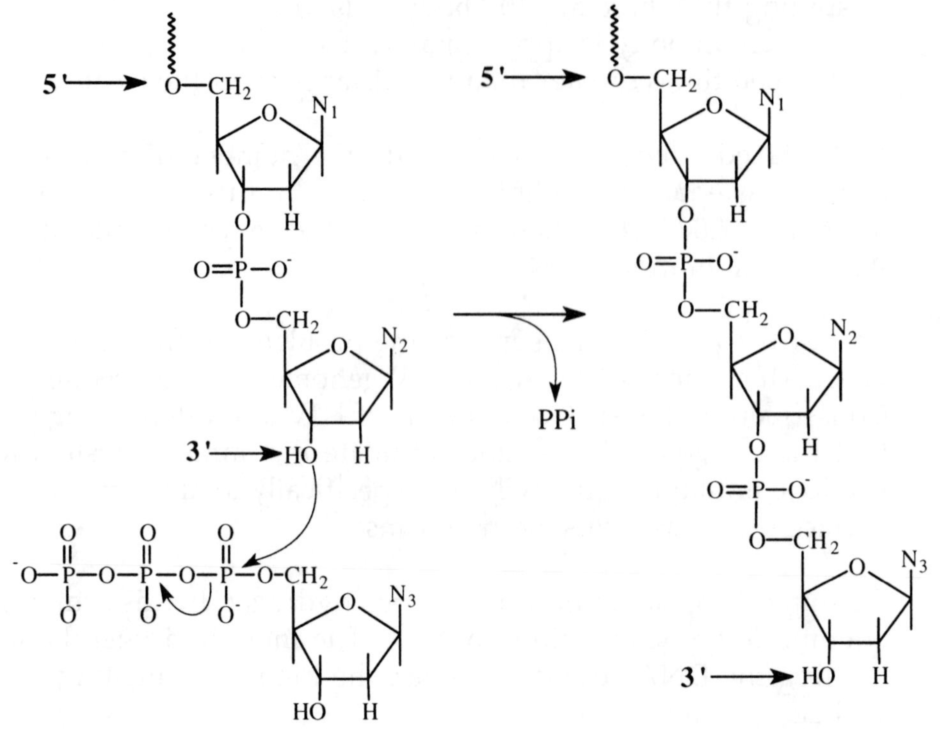

5' → O—CH₂ ... N₁ ... O H ... O=P—O⁻ ... O—CH₂ ... N₂ ... 3' → HO H ... ⁻O—P—O—P—O—P—O-CH₂ ... N₃ ... HO H → PPi

5' → O—CH₂ ... N₁ ... O H ... O=P—O⁻ ... O—CH₂ ... N₂ ... O H ... O=P—O⁻ ... O—CH₂ ... N₃ ... 3' → HO H

10. a. ROS may cause single and double strand breaks, pyrimidine-dimers, and the loss of purine and pyrimidine bases.

b. Because caffeine is a base analogue of thymine, it can cause transition mutations.

c. Small alkylating agents would attach to the nitrogen atoms of the purines and pyrimidines, destabilizing glycosidic linkages and interfering with hydrogen bonding and promoting both transversion and transition mutations.

d. Large alkylating agents would have the same effects as small alkylating agents but in addition they could also behave similarly to intercalating agents leading to frame shift mutations and breakage of the DNA chain.

e. Nitrous acid deaminates bases. For example, cytosine is converted to uracil.

f. Intercalating agents cause deletion or insertion mutations.

12. In excision repair short damaged sequences (e.g., thymine dimers) are excised and replaced with correct sequences. After an endonuclease deletes the damaged single-stranded sequence, a DNA polymerase activity synthesizes a replacement sequence using the undamaged strand as a template. In photoreactivation repair, a photoreactivating enzyme uses light energy to repair pyrimidine dimers. In recombinational repair, damaged sequences are deleted.

Repair involves an exchange of an appropriate segment of the homologous DNA molecule.

14. Because of the complexity of even the smallest genomes, most changes in DNA sequence have the potential to disrupt to a greater or lesser extent the smooth functioning of the organism.

16. Genetic recombination can allow cells to alter gene expression. The best known example is antibody production in lymphocytes. The rearrangement of several possible choices for each of a number of antibody gene segments via site specific recombination results in the generation of an extremely large number of different antibody molecules.

18. Mustard gas, a bifunctional alkylating agent, inhibits replication by crosslinking DNA strands.

20. Any exposure to UV light causes genetic damage in skin cells. The production of melanin, the "tanning" substance that absorbs the energy of UV light, is a response to damage that has already occurred. This damage, which accumulates over years of exposure to UV light, accelerates the aging process, hence the wrinkled and thickened skin. In some genetically predisposed individuals, the accumulating damage results in skin cancer.

22. Recall that phorbol esters mimic the action of diacylglycerol (DAG), the normal cell metabolite that activates protein kinase C (PKC). PKC initiates a phosphorylation cascade that results in the activation of numerous molecules involved in cell growth and division, including jun and fos which then combine to form AP-1. AP-1 is a transcription factor that promotes cell division. Its formation causes an affected cell to have a growth advantage over nearby cells. Because phorbol esters are tumor promoters, any exposure to them increases the risk that initiated cells may progress towards a cancerous state.

24. Because the Rb gene codes for a tumor suppressor, retinoblastoma only occurs when both copies have been damaged or deleted. Usually a long period of time is required for random mutations to cause this event. In hereditary retinoblastoma, in which an affected individual possesses only one functional Rb gene, the time span necessary for a random mutation to inactivate the second Rb gene is significantly less than that required for the inactivation of both genes that cause the nonhereditary version of the disease.

Chapter 18 Answers to Even Numbered Questions

2.   According to the RNA world hypothesis, in the evolution of life on earth, RNA molecules appeared before either DNA or protein. The hypothesis assumes that RNA molecules, which possessed both catalytic and informational properties, were capable of some degree of self-replication and possessed a capacity to catalyze primitive reactions.

4.   The observations upon which the wobble hypothesis is based are: (1) the first two base pairings in a codon-anticodon interaction confer most of the specificity required during translation, and (2) the interactions between the third codon and anticodon nucleotides are less stringent. Because of the "wobble rules," only a minimum of 31 tRNAs are required for the translation of all 61 codons.

6.   The sequential reactions that occur within the active site of aminoacyl-tRNA synthesis are: (1) the formation of aminoacyl-AMP, which contains a high energy mixed anhydride bond, and (2) linkage of the aminoacyl group to its specific tRNA.

8.   The three phases of protein synthesis are: (1) initiation (the small ribosomal subunit binds to an mRNA, an initiation tRNA and the large subunit), (2) elongation (polypeptide synthesis occurs as amino acids attached to tRNAs aligned in the P and A sites undergo transpeptidation), and (3) termination (polypeptide synthesis ends when a stop codon on the mRNA enters the A site). Translation factors perform a variety of roles. Some have catalytic functions (e.g., EF-TU and eEF-2 are GTP binding proteins which catalyze GTP hydrolysis), while others (e.g., IF-3 and eIF-4F) stabilize translation structures.

10.   The phases of protein synthesis during which each process occurs are as follows: a. Initiation, b. Elongation, c. Elongation, and d. Termination.

12.   The nucleotide GTP is the source of the energy required to drive various steps in the translation mechanism.

14.   Posttranslational modification reactions prepare polypeptides to serve their specific functions and direct them to specific cellular or extracellular locations. Examples of these modifications include proteolytic processing (e.g., removal of signal proteins), glycosylation, methylation, phosphorylation, hydroxylation, lipophilic modifications (e.g., N-myristoylation and prenylation) and disulfide bond formation.

16. During the elongation phase of protein synthesis, the second aminoacyl-RNA becomes bound to the ribosome in the A site. Peptide bond formation is then catalyzed by peptidyl transferase. Subsequently, the ribosome is moved along the mRNA by a mechanism referred to as translocation.

18. Kinetic proofreading is a mechanism that insures the correct codon-anticodon pairing in the A site of ribosomes. In eukaryotes eEF-1α mediates the binding of aminoacyl-tRNAs to the A site. When the correct pairing occurs eEF-1α hydrolyses its bound GTP and subsequently exits the ribosome. If correct pairing does not occur, the eEF-1α-GTP-aminoacyl complex leaves the A site, thereby preventing the incorporation of incorrect amino acids.

20. Because of the degenerate nature of the genetic code, there are a large number of possible mRNA base sequences that might code for the amino acid sequence of each polypeptide.

22. Cotranslational transfer is a process in which nascent polypeptides are inserted through an intracellular membrane during ongoing protein synthesis. An integral membrane protein complex referred to as the translocon, mediates the transfer of polypeptides (each of which contain some hydrophilic residues) across the hydrophobic core of the membrane.

24. Sets of amino acids which may require proofreading include phenylalanine/tyrosine, serine/threonine, aspartate/glutamate, asparagine/glutamine, isoleucine/leucine, and glycine/alanine.

26. Synthesis of a secretory glycoprotein begins on a ribosome. An appropriate signal peptide mediates the translocation of the polypeptide into the ER lumen. The core N-linked oligosaccharides are then covalently linked to appropriate asparagine residues in the polypeptide in a reaction catalyzed by glucosyl transferase. Subsequently, the molecule is transferred in transport vesicles to the Golgi complex where additional glycosylation reactions occur. Eventually, the glycoprotein is incorporated into secretory vesicles which migrate to the plasma membrane. Secretion of the glycoprotein then occurs via exocytosis.

28. The hsp70s are a family of ATP-binding molecular chaperones that bind to and stabilize protein during the early stages of protein folding. The hsp60s form a large structure composed of two stacked rings in which proteins fold in an ATP-dependent process.

Chapter 19 Answers to Even Numbered Questions

2.   Methods employed by living organisms that protect against predation include the production of antibiotics and toxins that eliminate the predator. Obnoxious tasting or smelling molecules also protect the organisms that produce them.

4.   During periods of environmental change, random genetic variations that arise from mutations and recombination may inadvertently confer a survival and/or reproductive advantage on the organism in which they occur. The progeny of this organism survive and depending on the level of selection pressure, may become the dominant strain in the population. One of the best documented examples of this phenomenon is the spread of antibiotic resistance in bacterial populations.

6.   a.   Examples of alkaloid secondary metabolites include atropine and nicotine.
      b.   Examples of terpene secondary metabolites include the carotenoids, vitamin K and camphor.
      c.   Examples of glycoside secondary metabolites include ouabain and digitoxin.

8.   During phase I of biotransformation, hydrophobic substances are converted into more polar molecules by reactions catalyzed by oxidoreductases and hydrolases. During phase II of biotransformation, the solubility properties of a molecule are improved when appropriate functional groups are conjugated with substances such as glucuronate, sulfate, glutamate, or glutathione.

10.   The cytochrome P_{450} system is composed of two enzymes: NADPH-cytochrome P_{450} reductase and cytochrome P_{450}. Cytochrome P_{450} catalyzes oxygenation reactions. The two electrons required in each of these reactions are transferred one at time from NADPH by NADPH-cytochrome P_{450} reductase.

12.   The oxygenation of benzene is initiated when the substrate binds to oxidized cytochrome P_{450} (Fe^{+3}). This binding promotes a reduction of the enzyme substrate complex by an electron transferred from NADPH via cytochrome P_{450} reductase. After reduction, cytochrome P_{450} binds O_2. A second electron transferred from cytochrome P_{450} reductase results in the generation of a Fe^{+3}-O_2^--benzene complex. The oxygen-oxygen bond is broken with one atom released as water, while the other remains bound to heme. Abstraction of the hydrogen of benzene is followed by the transfer of the oxygen species from heme to the substrate. The product of the

212

reaction is an epoxide which is subsequently converted to phenol by epoxide hydrolase.

14. Flavin-containing monooxygenases differ from all other mammalian monooxygenases in that the presence of substrate is not required for oxygen reduction. A consequence of this phenomenon is that substrate specificity is broad. i.e., virtually any type of molecule can be oxygenated.

16. Commonly used donor substrates in phase II conjugation reactions include glucuronic acid, glutathione, sulfate, and amino acids.

18. The CNS effects of cocaine are apparently caused by blockage of the reuptake of neurotransmitters such as norepinephrine, dopamine and serotonin at nerve terminals. Most cocaine molecules are inactivated by esterases in blood plasma. The remaining cocaine molecules are oxidized by microsomal enzymes, primarily those of the liver. Liver damage often occurs in cocaine users, primarily because of the formation of the norcocaine nitroxide, the free radical product of the liver's oxidative processing of cocaine.

20. The epoxide product of tetrachloroethylene oxidation is:

$$\begin{array}{c} Cl \diagdown \quad \diagup Cl \\ C - C \\ Cl \diagup \; \diagdown O \diagup \; \diagdown Cl \end{array}$$

Liver damage is presumably caused by the reaction of the epoxide with DNA, RNA and proteins in the hepatocytes in which tetrachloroethylene is biotransformed.

22. If both rifampicin and an oral contraceptive are metabolized by the same cytochrome P_{450} system, the oral contraceptive may be metabolized so quickly because of the induction properties of rifampicin that its concentration in the body becomes ineffective.

24. Ethanol is toxic to the liver because its detoxification in that organ results in the formation of acetaldehyde, a molecule that forms covalent bonds with proteins and promotes lipid peroxidation. Because ethanol-inducible cytochrome P_{450} can convert certain xenobiotics to toxic metabolites, the use of tobacco in combination with excessive alcohol consumption significantly increases the risk of certain types of cancer. Recall that tobacco smoke contains a wide variety of carcinogens and tumor promoters.

26. Cross tolerance occurs when a xenobiotic that induces increased synthesis of a detoxicating mechanism (e.g., a cytochrome P_{450} system) is consumed simultaneously with another substance that is also biotransformed by the same mechanism. Soon the second drug requires an increased dosage for the maintenance of its clinical effectiveness because its metabolism is accelerated by the induction process.